PRELUDE TO CHEMISTRY

THE M.I.T. PAPERBACK SERIES

PRELUDE
TO CHEMISTRY

AN OUTLINE OF ALCHEMY
Its Literature and Relationships

by
JOHN READ

THE M.I.T. PRESS
Massachusetts Institute of Technology
Cambridge, Massachusetts, and London, England

First published in 1936
by G. Bell and Sons Ltd, London, England
Second edition 1939
Reprinted 1961
First M.I.T. Press Paperback Edition, August 1966

Library of Congress Catalog Card Number: 66-23892
Printed in the United States of America

TO MY WIFE

Whilom ther was in a small vilage,
As my Auctor maketh rehersal;
A *Chorle* the wich had lust and gret corage,
Within hymselfe by hys deligent travel,
To aray hys Garden with notabil reparel:
  Of lenght and brede y lyche square and long,
  Heggyd and dychyd to make yt sure and strong.

All the Aleys made playne with Sande,
Benches coverid with new Turves grene,
Set Erbes with Condites at the ende;
That wellid up agen the Sun schene,
Lyke Silver stremys as any cristal clene:
  The burbely Waves up ther on boylnyg,
  Rownde as Beral theyr bemys owte chedyng.

Mides the Garden stode a fresh Lawrer,
Ther on a *Byrde* syngyng both day and nyght;
With shinyng federis brighter then Gold weer,
Wych wyth hir song made hevy hertis lyght;
For to behold hit was an hevenly syght:
  How towerd evyn and in the dawnyng,
  Sche dyd her payne most ameus to syng.

Hermes Bird

CONTENTS

ix

PRELUDE TO CHEMISTRY

CHAPTER III—(*continued*)

CHAPTER IV

CHAPTER V

CHAPTER VI

CONTENTS

CHAPTER VII

ILLUSTRATIONS IN THE TEXT

PRELUDE TO CHEMISTRY

PLATES

PLATES

xvii

PRELUDE TO CHEMISTRY

THE PROHEME

HIS book has grown naturally out of occasional lectures of a historical kind which the author has delivered to general audiences. It does not profess to add to the formal literature of historical chemistry, of which assuredly there is no lack; nor does it provide a cut-and-dried account, in exhaustive detail, of the origin and development of chemistry. Its aim is rather to offer a bird's-eye view of chemistry's precursor, alchemy, followed by closer glimpses of certain fields of that richly coloured panorama of the ages. ' Thus with imagined wing our swift scene flies ' from the temples of ancient China and Egypt to the gloomy cells of the ' puffers ', the cities, mines and monasteries of mediaeval Europe, and the laboratories of Boyle and Lavoisier.

Many and romantic were the figures which flitted across the alchemical stage in the course of its fifteen centuries, or more, of existence. We see their dim reflections in our chymic mirror, caught up not so much from Thomson, Kopp, Hoefer, Ferguson, and others of that elect brotherhood of the chemical historians, as from the original manuscripts and printed tomes which lie neglected on the shelves of certain libraries and appear furtively at times in the bookshop and auction-room: for, as Elias Ashmole wrote in the preface to his *Theatrum Chemicum Britannicum* in 1652, ' *Old* words *have strong* Emphasis; *others may look upon them as* Rubbish *or* Trifles, *but they are grosly* Mistaken: *And* Posterity *will pay us in our own* Coyne, *should we deride the* behaviour *and* dresse *of our* Ancestors. *And therefore that the* Truth *and* Worth *of their* Workes *might receive no* Diminution *by my* Transcription, *I purposely* retain'd *the old* Words *and manner of their* Spelling, *as I found them in the* Originalls.'

Among the alchemical figures in our chance selection are some whose very names will ring strangely in the ears of modern students and practitioners of chemistry. Here we shall catch fleeting images of Nicolas Flamel and his devoted Perrenelle, whom he loved as himself; of Black Berthold, the experimenting monk of Freiburg im Breisgau; of Agricola, and his sprites of the mine; of Andreas Libavius, Alexander Seton, Heinrich Khunrath, rare Ben Jonson, and many another. In our closing scenes the dominating actors are the mysterious Abbot Cremer of Westminster; the still more enigmatical Basil Valentine, the 'mighty king' of alchemy; Tomais Norton of Briseto ('a parfet Master ye maie him call trowe'); Michael Maier, Count of the Imperial Consistory, Doctor of Physic and Medicine, sometime Physician to the Emperor Rudolph II, alchemist, philosopher, mystic, classical scholar, and amateur of music; Master John Daniel Mylius, that industrious Chymist; the most accomplished and cultured Master Lucas Jennis, the prosperous Frankfurt publisher and patron of alchemy; and last, but of a truth not least, Daniel Stolcius, the modest and unknown student of poetry and medicine, longing, in a self-imposed exile, for his native Bohemia.

In a sense, it is to Daniel Stolcius that this book is due. About seven years ago, a copy of his extremely rare *Viridarium Chymicum*, or ' Pleasure Garden of Chymistry ', a work of peculiar interest and charm, fell, with the unexpectedness of Basil Valentine's thunderbolt at Erfurt, into the hands of the present writer. The last four chapters of this book form an immediate outcome of the work of interpreting and tracing the origins of the 107 emblematic engravings contained in Stolcius' fascinating little volume.

THE PROHEME

Alchemy was pictorial in its expression to a degree which is not realised in this age, and that is why this new *Viridarium* is so fully illustrated.   The woodcuts and copper-plate engravings which adorn so many al-chemical treatises bear eloquent testimony to the artistic talent and skilled craftsmanship of the early workers in these alluring media.   Like much of the early printing, they carry the impress of fresh minds and eager hands upon a new art.   Pictorial alchemy offers indeed many attractions, and possibly some lessons, to the modern chemist who glories in the compact expressiveness of the chemical formula: how much more inspiring, for ex-ample, is the pictorial idea of the combination of the ' two sulphurs ', shown in the guise of the frisky lions of Plate 57, than some such prosaic formulation as $Hg + S = Au$!   There are those, as Poisson remarks, to whom such symbols are meaningless figures drawn at random; but it is easy to despise something which one makes no effort to understand.

It is sometimes said that an increasing tendency of modern science to exalt the importance of ' learning more and more about less and less ' is likely to produce a con-dition of intellectual myopia in the scientific specialist. If this be so, it is difficult to imagine a more stimulating and picturesque corrective than the development of an interest in historical science.   ' It is only when science is explained and tempered by history that it acquires its whole educational value, and that the main objections to scientific education cease to be valid.   The more science enters into our lives, the more it must be " humanised ", and there is no better way to humanise it than to study its history.   Such studies, reconciling the purely scientific, the historic, and the philosophic points of view, would be the source of the soundest and highest idealism.' [1]

The amazing width of horizon which studies of this kind can disclose is apparent from the chapters which follow.   Here, chemistry is seen to possess numerous links of absorbing interest with history, literature, mythology, astrology, folk-lore, mysticism, philosophy, religion, pictorial art, and even music.   Should this publication help in any way to bring refreshment to ' wearied Servants of Laboratories ', the writer will be encouraged to pursue his original design of filling in the outlines of a further sequence of alchemical and chemical scenes, dealing partly with later times, for which there is no room in the present collection.

The reproduction in this work of so many typical alchemical illustrations, selected mainly from rare books and manuscripts, has only been rendered possible through the application to this purpose of a generous grant from the Carnegie Trust for the Universities of Scotland: grateful acknowledgment is now made for the invaluable help received from this source.

The author gladly records his indebtedness to various corporations and societies for allowing him to draw fully upon the subject-matter of the lectures from which this volume has grown: thanks are due especially to the Folk-Lore Society, the Royal Institution of Great Britain, and the Thomas Young Club of Emmanuel College, Cambridge.   It is a pleasure also to acknowledge the help that has been derived from certain transatlantic authors whose scholarly and original work forms a striking feature of the recent revival of interest in historical science. In this place, particular reference is due to two of them. Professor Tenney L. Davis, of the Massachusetts Institute of Technology, to whom the writer is indebted for kind permission to reproduce Plates 1 and 20, has made additions of the first importance to our knowledge of Chinese alchemy and to the problem of alchemical origins; and Professor Arthur J. Hopkins, of Amherst College, Massachusetts, whose ' colour theory ' of the origin of alchemy deserves a wider consideration than it has yet received, has rendered valuable service to historical chemistry in his studies of Greek alchemy.

Most of the books, and one or two of the manuscripts, which are considered in detail below find a place in the

library of the Chemistry Department of the United
College of St. Salvator and St. Leonard, in the University
of St. Andrews. In addition, much help has been de-
rived from other sources, and the writer acknowledges
in particular the kind services of: the Keeper of the
Department of Manuscripts, British Museum; Dr. Josef
Volf, Director of the Library of the Národní Museum,
Prague; M. F. de Montremy, Musée de Cluny, Paris;
Dr. W. Douglas Simpson, University Librarian,
Aberdeen; Mr W. G. Burrell, Librarian of the Royal
Technical College, Glasgow; Professor T. S. Patterson,
University of Glasgow; Professor Dr. J. V. Dubský,
Masarykovy University, Brno; Mr H. C. Hoover,
Stanford University, California; and Principal Sir J. C.
Irvine, University of St. Andrews. Thanks are due also
to the Librarian of the University of Glasgow for kind
permission to photograph and reproduce the title-page of
the copy of Maier's *Viatorium* (1618) in the Ferguson
Collection (Plate 13).

In conclusion, it must be stated that the production
of this book in its present form would have been impossible without the generous help, so cordially given,
of two of the author's colleagues in the University of
St. Andrews. Mr F. H. Sawyer, M.A., B.Mus., and
Professor H. J. Rose, M.A., F.B.A., have placed their
specialised knowledge freely at the author's disposal in
those parts of the work in which alchemy is linked on the
one hand with music and on the other with mediaeval
Latin and classical mythology. It is of interest to mention that examples of Count Michael Maier's alchemical
' fugues ' (p. 281) were sung under Mr Sawyer's direction
at the Royal Institution of Great Britain on November
22nd, 1935, by student members of the St. Andrews
University Choir (informally known as ' The Chymic
Choir '), this being in all probability the first performance in public of any alchemical music. Four of
Maier's 'fugues', together with a modern chemical canon
composed by Professor Rose and Mr Sawyer under the

title 'Hydrogenesis', all of which were sung upon that occasion, are reproduced in an Appendix to this work. Professor Rose's felicitous translations of Latin passages, often of considerable obscurity, and Mr Sawyer's annotated transcriptions of Maier's alchemical music are truly 'notabil reparel' for this modern Pleasure Garden of Chemistry.

The Garden has been arrayed to lure *Bees* and *Butterflies* alike; but it may be added that a flowery Border, metamorphosed after the manner of alchemy into a bibliography, garnishes the outgoing Arbour. Moreover, divers faire Rose-trees, transmuted into cross-references, have been planted in the middest of the Garden. The *Butterflies*, an they list, may hover over these Blossoms without settling thereon; but peradventure the *Bees* will find them as full of Honey as of Fragrance.

J. R.

THE UNIVERSITY,
  ST. ANDREWS, SCOTLAND

Her ys an Erbe men calls *Lunayrie*,
I-blesset mowte hys maker bee.

CHAPTER I

AN OUTLINE OF ALCHEMY

*The systems which confront the intelligence remain basically un-
changed through the ages, although they assume different forms . . .
there is nothing so disastrous in science as the arrogant dogmatism
which despises the past and admires nothing but the present.*

                                                            HOEFER

*In chemistry also, we are now conscious of the continuity of man's
intellectual effort; no longer does the current generation view the
work of its forerunners with a disdainful lack of appreciation; and
far from claiming infallibility, each successive age recognises the
duty of developing its heritage from the past.*   KEKULÉ

§ 1. NATURE AND ORIGINS

 ANY writers have interpreted alchemy in a
restricted sense, as the pretended art of
transmuting the so-called ' base ' metals
into the ' noble ' metals, silver and gold.
This point of view is not confined to
modern times, for long ago John Gower [1]
wrote of ' thilke Experience which cleped
is *Alconomie*, Whereof the Silver multeplie; Thei made,
and eke the Gold also '. Much later in the era of alchemy,
Salmon stated in his *Polygraphice*,[2] which had so great
a vogue in the late seventeenth century, that ' *Alchymie*
is an *Arabick* word, and signifies the Transmutation of
Metals, Semi-Metals and Minerals '.

In a wider sense, alchemy was the chemistry of the
Middle Ages: according to Liebig,[3] indeed, ' Alchemy
was never at any time anything different from chemistry.
It is utterly unjust to confound it, as is generally done,
with the gold-making of the sixteenth and seventeenth

centuries. . . . Alchemy was a science, and included all those processes in which chemistry was technically applied.'

In its broadest aspect, alchemy appears as a system of philosophy which claimed to penetrate the mystery of life as well as the formation of inanimate substances. A modern writer on this wider alchemy voices some of its claims in the following language: [4] ' Hermetism, or its synonym Alchemy, was in its primary intention and office the philosophic and exact science of the regeneration of the human soul from its present sense-immersed state into the perfection and nobility of that divine condition in which it was originally created. Secondarily and incidentally . . . it carried with it a knowledge of the way in which the life-essence of things belonging to the sub-human kingdoms—the metallic *genera* in particular—can, correspondingly, be intensified and raised to a nobler form than that in which it exists in its present natural state.'

Another modern exponent of alchemy states: [5] ' The operation of the Philosopher's Stone does not belong to the realm of pure chemistry. The method described with so remarkable a unity of doctrine excludes any idea of research or tentative procedure, and is incompatible with the abundant experimentation involved in modern chemistry, both organic and inorganic. In this method we see the working of an eagerness, an inspiration, and a fertilising and generative element showing that the alchemists had surprised some secret of cellular life which, carried into the metallurgic field, produced effects unknown now. . . . It is not then from alchemy, as often stated, that modern chemistry derives, but actually from the erratic work of the Puffers.' As this extract indicates, a distinction is often drawn in alchemical writings between an esoteric alchemy, whose hidden secrets were revealed only to chosen adepts, and an exoteric pseudo-alchemy, which is depicted as the uninstructed craft of mercenary gold-seekers, or ' puffers '.

AN OUTLINE OF ALCHEMY

All modern pronouncements on alchemy are necessarily the opinions of experts, for in the twentieth century 'the Divine Art' has lost the wide appeal which it possessed three hundred years ago. To-day, the man-in-the-street, so far as he is cognisant of alchemy at all, pictures it merely as a collection of old wives' tales arising from those crazy cauldrons which he regards as indispensable stage-properties of the comic relief of the Middle Ages.

It is interesting to add to these divers views of the nature of alchemy the opening words of Lenglet Dufresnoy's *Histoire de la Philosophie Hermétique*,[6] published in 1742: 'I am about to give in this little work the history of the greatest folly, and of the greatest wisdom, of which men are capable'.

Faced with this diversity of opinion, we may well ask: What, then, was alchemy? How did it arise? What were the ideas which influenced the alchemists? What was the driving force which actuated them? What did they accomplish?

In order to find answers to such questions as these, let us go forth—like the 'alchemystical philosopher' of *Splendor Solis* (p. 67)—in search of the elements of alchemy. An acquaintance with chemistry will help us in our quest; but even more valuable will be that combination of general knowledge and common sense, that capacity to weigh up evidence, that appreciation of the humanistic appeal of chemistry, and that power of seeing the wood despite the trees, which all good chemists ought to possess.

Armed with such simple weapons, let us fare back from modern chemistry into the old forgotten world of alchemy. Let us try to assimilate something of the alchemistic inheritance, environment, and point of view. Let us examine for ourselves, giving at the same time due heed to the commentators, some of the cryptic writings of the 'sons of Hermes'; and let us seek out the hidden meanings of some of their intriguing symbols and enigmas.

3

The task is a fascinating one; and although we may not always—in alchemical phrase—' hitt the *Marke* ', we shall surely gain from this pleasant recreation a shrewd idea of the modes of thought and work of these far-away practitioners of ' the Divine Art ', who were our spiritual ancestors.

Scientific chemistry has been called the outcome of man's attempts to produce gold artificially and to explain its occurrence in the earth's crust,[7] and the origin of alchemy has often been associated with early evidences of the importance attached to gold. In China, and also in India, from remote times, gold was valued as a magic medicine. In Egypt, there were expert goldsmiths as early as 3000 B.C., and in the Euphrates valley skilled Sumerian metal-workers practised their craft some five hundred years earlier still. Such processes as the fusion and weighing of gold in ancient Egypt were depicted on tomb walls, dating, as at Beni Hassan, from about 2500 B.C. The oldest map in the world, which was drawn more than three thousand years ago (*c.* 1300 B.C.), depicts a gold-mining region in the eastern desert of Egypt. The lavish use which was made of gold in Egypt at that time may be inferred from the contents of the tomb of Tut-ankhamûn (1350 B.C.): the solid gold coffin of this ' nonentity among pharaohs ' weighed more than two hundredweight (100 kilos).[8]

There is ample evidence that the Egyptians were remarkably skilled in various arts based upon chemical knowledge, such as metallurgy, enamelling, glass-tinting, the extraction of plant oils, and dyeing. For such reasons, Egypt, or Khem,[9] the country of dark soil, the Hebrew ' Land of Ham ', has often been pictured as the motherland of chemistry; so that later this ' art of the dark country ' became known to Islam as *al Khem*, and through Islam to the Western world as *alchemy*.

The reputed Egyptian origin of alchemy is encour-aged throughout the vast body of mediaeval alchemical literature by constant references to Hermes Trismegistos, or Hermes the Thrice-Great, the alleged father of the 'Hermetic Art' and the patron of its practitioners, the self-styled 'sons of Hermes'. Hermes is regarded as the Greek equivalent of the ibis-headed moon-god, Thoth, who, in turn, has sometimes been identified with the deified Athothis of 3400 B.C., or with Imhotep, both of whom excelled in the art of healing. Thoth was the Egyptian god of healing, of intelligence, and of letters: in the last capacity, his Greek equivalent, Hermes, may perhaps be excused for the 36,000 original writings which he was said to have contributed to alchemical literature! Of these, the best known is the so-called Emerald Table of Hermes, consisting of a sequence of sentences or precepts, which are frequently quoted in alchemical writings (p. 54). Some of the characteristics attributed to the Greek god Hermes are of interest in alchemy. Occasionally he is pictured as a new-born babe. He was sometimes regarded as the god of fertility, and as a patron of music. The sacred number 4 was assigned to him. As the messenger of the gods, Hermes is endowed with mobility; he wears winged boots and carries a caduceus, or herald's staff, which is shown as a winged wand entwined by two serpents. The caduceus is a familiar emblem in alchemy (p. 106), and considerable symbolic significance has been associated with its design.

In recent years Egypt has found a rival in China, as the possible original home of alchemy. Alchemical ideas appear to have arisen in China as early as the fifth century B.C., and alchemy is said to have been actively pursued in that country from 300 B.C. onwards.[10] Chinese alchemy is closely bound up with Taoism, a system of philosophy and religion, which was based partly on the teachings of Lao-Tsŭ of the sixth century B.C. and given form and direction by Chang Tao-ling in the second century A.D. The Taoist Canon is a rich repository of ancient alchemistic writings, which await detailed study and correlation. According to Waley,[11] the ancient Chinese had inherited from remote ages a belief in the life-giving and rejuvenating properties of jade, pearl, cinnabar, and other materials. In extant alchemical treatises dating from about 317–332 A.D., Ko Hung (Plate 1), writing under the philosophical pseudonym of Pao P'u Tzŭ,[12] discriminates between three kinds of alchemy: (1) the preparation of a liquid gold, producing longevity; (2) the production of artificial cinnabar, the ' life-giving ' red pigment, for use in gold-making; (3) transmutation of base metals into gold. In the tenth century, exoteric alchemy (*wai tan*) merged into esoteric (*nei tan*), according to which the ' souls ' or ' essences ' of mercury, sulphur, lead, etc., bear to the common metals the same relationship as the Taoist, adept, or perfected man, bears to common mortals. A further evolution had been reached at the end of the eleventh century, by which time the transcendental metals had become identified with various parts of the human body.

These leading ideas are so strongly reminiscent of the tenets of Western alchemy that a common origin has been sought for the two schools of thought. Some authorities hold that alchemistic ideas filtered into China from Egypt, Mesopotamia, or India, in the second or third century B.C.; and, according to von Lippmann, alchemy proper reached China from the West in the eighth century A.D. through the medium of Arab trading ships. On the other hand, it has been suggested that alchemy arose first in China, from the philosophy of Taoism, and then spread to the West ; in particular, there is much evidence in support of a Chinese derivation for the idea of the elixir of life.[13] A similar idea is encountered in the *Arthavaveda* of the Hindus, dating from 1000 B.C. or earlier: [14] the Hindus, like the Chinese, associated medicinal gold with longevity and even with immortality.

Davis states that, because of the imagined power of artificial gold to produce longevity, the Chinese attempted to transmute base metals into gold as early as the second and third centuries B.C.; he has adduced impressive evidence [15] in favour of an Oriental origin of alchemy, which he regards as a development of Taoist mysticism in the third or fourth century B.C. Lu-Chiang Wu's translations of the *Ts'an T'ung Ch'i* [16] and portions of *Pao-p'u-tzŭ* [15] furnish evidence—to quote Davis' own words—' that European alchemy was derived from that of China, that Chinese alchemy reached the Arabs, and thence Europe, through Persia, with which country the Chinese had intercourse both before and after its conquest by ·the Mohammedans. There is no evidence that alchemy existed in Europe, or in Byzantium or Alexandria, before the eighth century or thereabouts when the Arabs began to practise it — and the opinion . . . that Chinese alchemy came to Europe through Byzantium and Alexandria cannot be accepted. Indeed the present translation [*Pao-p'u-tzŭ*] supplies new evidence of the fundamental dissimilarity between the aims of the Chinese alchemists who sought to make real gold and silver artificially and those of the Alexandrian and Byzantine chemists who strove to tincture the base metals to the appearance of the noble ones.' Furthermore, it is significant that the doctrine of the Two Contraries (p. 19), which appears to form the ultimate basis of alchemical theories, was widely diffused before Greek philosophy arose.

The ultimate origin of alchemy is thus a vexed question: the claims of the Orient have been strengthened by the researches of the twentieth century, and in any case it appears that China possesses, in the *Ts'an T'ung Ch'i* of Wei Po-yang (p. 38), the earliest known treatise (*c.* 142 A.D.) which is devoted entirely to alchemy.[16] It is indeed a baffling task to unravel after the lapse of so long a period the many threads which were woven at different times and in various countries into the fabric of early alchemy. In considering this problem, it must be realised that there was perhaps a much fuller contact between the civilisations of the ancient world than is sometimes imagined, and that the task of assigning origins to a specific country or civilisation is correspondingly difficult. Partington,[17] for example, has shown in detail how the technical arts of Greece and Rome were based upon knowledge derived from the much older cultures of Egypt, Babylonia and Assyria, the Ægean, Asia Minor, Persia, Syria, Palestine, etc. This derived knowledge of materials and their uses, he points out, often represents ' not an original and vigorous development of national genius, but a decadent form of craftsmanship which had existed for a period often as long as that which now separates us from the best days of Greece and Rome '. Greek knowledge and theories, gathered from so wide a field, were in turn transmitted to Islam, mainly through Syria and Persia. That Islam's heritage of alchemical knowledge included bequests from Egypt, from Babylonia, the home of astrology and magic in the ancient world, from India, and from China, may be taken for granted. Eventually the accumulated knowledge of the Muslim chemists, brought together from these varied sources, was disseminated throughout western Europe, chiefly by way of Spain.

§ 2. THE GRAECO-EGYPTIAN ERA

From the evidence available, it is impracticable to assign the origin of alchemy to a definite place or time. Quite apart from ultimate origins, however, there is no doubt that the incipient art was influenced during the Alexandrian age (4th century B.C. to 7th century A.D.) by the application of Greek philosophy to the technique of the Egyptian and other ancient cultures. Holmyard [18] deals with this aspect of alchemy in the following words:

' Although there are dissentients, it is commonly believed that chemistry arose in the early years of the Christian era, as a result of the fusion of Egyptian metallurgical and other arts with the mystical philosophies of the Neo-Platonists and Gnostics. Unluckily, the Neo-

8

PLATE I

Ko Hung (*c*. 281–361 A.D.), Taoist philosopher, alchemist,
medical writer, and poet.

PLATE 2

(i) Maria the Jewess.

From *Viridarium Chymicum*, Stolcius, 1624 (reprinted from *Symbola Aureae Mensae Duodecim Nationum*, Maier, 1617). (See pp. 15, 258.)

(ii) Avicenna the Arabian.

From the same. (See pp. 17, 258.)

Platonists regarded matter as the principle of unreality
or evil, from which the disciple should attempt to detach
himself, while the Gnostics cared little for the phenomena
of the sensible world, being much more anxious to attain
to a knowledge of the invisible cosmos.   It is significant
for the later history of the science that one of the earliest
chemical writers, Zosimos the Panopolitan, was a Gnostic,
while the Neo-Platonic conceptions of sympathetic action,
action at a distance, the distinction between occult and
manifest properties, the influence of the stars, and the
mystical powers of numbers, all permeate chemistry from
its beginnings at the time of Plotinus until the close of
the seventeenth century.   It would, indeed, scarcely be
going too far to say that some of these ideas are with us
still: nitrogen is manifestly inert but occultly active, and
the structure of the atom is ultimately a matter of the
relations between numbers.'

Alchemy, like modern science, had its theories, although these were often vague, ill-defined, and subject to
interpretations suiting the whims of the interpreter.   The
chief physical theory of alchemy was that of the Four
Elements, or ' simple bodies '.   This is usually ascribed
to Aristotle (c. 350 B.C.), to whom it descended through
Plato (c. 400 B.C.) from Empedocles (c. 450 B.C.); but
the fundamental idea is said to have been recognised in
both India and Egypt as early as 1500 B.C.   An almost
equal antiquity has been assigned to the Chinese conception of the Five Elements (p. 20). We encounter
here, indeed, ' one of those crude physical theories which
is enunciated and accepted by races the most diverse in
character, country, faith, destiny.   There is great one-
ness in the human mind in the matter of broad principles
in crude cosmical ideas.' [19]

In adopting Empedocles' doctrine of the four ' elements', Aristotle followed Plato in rejecting their supposed
elementary nature and in considering that they were interconvertible.   Briefly, Aristotle's theory postulated the
existence, as abstract entities, of four fundamental pro-

9

perties or qualities of bodies: these were the hot and the moist, with their contraries, the cold and the dry. The four material elements—earth, air, fire, and water—were pictured as originating by pair-wise conjunctions of the four elementary qualities, as indicated diagrammatically in Fig. 1. Of these two pairs of contraries, fire and water were endowed with the greater significance.

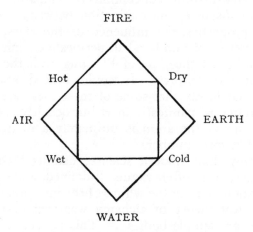

FIG. 1.—The Four Elements.

All bodies were held to be composed of the four elements in different proportions. As a corollary, one body could be changed, or transmuted, into another by altering the proportions of the elements present; further, this supposed transmutation was correlated with the idea of a *prima materia*, or primordial matter, from which all things came and to which they all reverted. An ' eternal principle ' was recognised also in ancient India under the name of *Brahma*,[20] and the ' primal matter ' of the ancient Chinese was known as *t'ai-chi*.

The theory may be illustrated by quoting a simple example of a supposed transmutation. Water is wet and cold; when the cold is expelled through the application of heat, the water changes into air (*i.e.* an invisible vapour), which is wet and hot. Earth, water, and air apparently typified the solid, liquid, and gaseous states of matter, while fire typified combustibility and possibly energy. As Muir [21] remarks: ' In former times all liquid substances were supposed to be liquid because they possessed something in common; this hypothetical something was called the *Element, Water*. Similarly, the view prevailed until comparatively recent times, that burning substances burn because of the presence in them of a hypothetical imponderable fluid, called " *Caloric* "; the alchemists preferred to call this indefinable something an Element, and to name it *Fire*.'

The system of four elements is often symbolised in alchemical writings as a square, and the individual members are given distinct triangular signs (p. 209): those of fire and air point upwards, those of water and earth downwards, and the symbols for air and earth are crossed by a bar.

Whatever the ultimate origin of alchemy may have been, there is no doubt that Greek thought and Greek ideas exerted a powerful influence upon mediaeval alchemy. Thus, the theory of the four elements entered into the philosophy of Plato and Aristotle, and many other alchemical ideas were drawn from the same source. Notably, the idea of matter playing the rôle of a passive recipient upon which qualities may be imposed, thus giving rise to metals and other ' forms ' of matter, is derived from Plato's *Timaeus*. Or again, the alchemists held that metals pass through a cycle of growth, culminating in the perfection of gold, and that in these changes art can assist Nature: this belief is clearly related to Aristotle's dictum that Nature strives towards perfection, and less directly to Plato's earlier pronouncement that nothing exists which is not good. The wide diffusion of Pythagorean ideas throughout mediaeval alchemy is discussed later (p. 247).

Hopkins [22] stresses the influence upon alchemy of four fundamental beliefs which were also widely held in ancient Greece. These were, in his words:

'(1) Hylozoism: all Nature is like man, alive and sensitive.

(2) That the great universe of sun and stars, the "macrocosm", is guided by the same laws which obtain on the earth, [and for] the "microcosm" [of man's body].

(3) Astrology: the stars influence and foretell the course of events on this earth.

(4) Animism: any event apparently spontaneous is really due to some personality—fairy, wood spirit, hobgoblin, etc.'

Still other influences were derived from the various religious beliefs of Alexandrian Egypt; particular importance is ascribed by Hopkins,[23] Davis,[24] and others to the belief, expressed in various ways, in a mediator acting between God and man.

The theory and practice of alchemy were strongly influenced for more than a thousand years by a belief in the existence of a potent transmuting agent, which ultimately came to be regarded as a universal medicine (p. 121). This is referred to in the writings of Western alchemists under many names, such as the Philosopher's Stone, the Elixir Vitae, the Grand Magisterium, Magistery, or Elixir; and the Red Tincture. Alchemical reasoning was mainly deductive reasoning, based upon two *a priori* assumptions: namely, the unity of matter and the existence of a Philosopher's Stone. It has already been said that the general idea of transmutation formed an integral part of the theory of the four elements. Beyond this, a belief in the possibility of a specific transmutation of base metals into silver and gold was no doubt fostered by a variety of experimental observations, such as the production of silver from the lead of galena and the preparation of alloys bearing a superficial resemblance to gold. The idea of the Philosopher's Stone may have arisen from the Alexandrian and Neo-Platonic belief in magic. It played a part also in a supposed analogy between the alchemical doctrine of transmutation and the mystical doctrine of the regeneration of man: it was the material analogue of that spirit of perfection which was able to cleanse the soul from the imperfections of sin.

The descriptions of the perfect Stone, or Elixir, were many and various. In some of them it was depicted as a heavy, glittering powder, which melted without producing smoke. The powder was red or white, in accordance with its power of converting the base metals into gold or silver when ' projected ' upon them. Van Helmont, in 1618, described it ' of colour, such as is in Saffron in its powder, yet weighty, and shining like unto powdered Glass '; Helvetius also, as late as 1667, said that the Tincture resembled glass or pale sulphur.

It has been suggested that the idea of transmutation originally arose from the discovery in ancient Egypt of alloys simulating gold, ' so that even the gold workers will be deceived ', as a reference runs in the Leyden papyrus of the first or second century A.D. (p. 38). From a study of the somewhat later writings of the Alexandrian alchemists (p. 40), Hopkins [25] reaches the conclusion that the original Egyptian ' gold-making ' and ' silver-making ' consisted merely in impressing the colours of gold and silver upon the baser metals, and that the mediaeval ' pseudo-alchemists ' of western Europe, by interpreting these terms literally, were led away on a false trail.

In a work of marked originality, entitled *Alchemy, Child of Greek Philosophy*,[26] this writer develops a plausible ' colour theory ' of alchemy which is worthy of close attention. ' In a land of neutral hues,' he remarks, ' colour appealed to the people as a delicious joy . . . colour hunger so filled the minds of the Egyptians that colour was . . . to them an ideal of perfection in the world of metal '. After some four thousand years' experience, the ancient Egyptians had acquired unexampled skill in the constructive and imitative arts. From the dyeing of fabrics they passed to the ' tinting ' (or ' tingeing ') of metals, for which they developed the processes of dipping into mordanting baths, alloying, and treatment with 'royal cement' and with 'sulphur water'. This last reagent was a solution of sulphuretted hydrogen, known also somewhat surprisingly as 'Holy Water', presumably on account of its mysterious odour, which was perhaps identified in ancient Egypt with the odour of sanctity. This potent weapon in the armoury of the original alchemists was made by heating sulphur with lime and pouring water on to the product: 'on opening the cover,' wrote Zosimos,[27] 'do not put your nose too close to the mouth of the jar'.

Zosimos, who lived in the third or fourth century A.D., like the later alchemists of the Middle Ages, seems to have regarded his 'tinted', imitative silver and gold as superior in colour and purity to the native metals. Since the tinted gold was yellower than ordinary gold, it was capable of imparting yellowness—or a yellow 'seed' or 'ferment'—to other metals, thus transmuting them into gold. This doctrine seems to lie behind the later alchemical idea of 'multiplication' by means of the Philosopher's Stone or Tincture.

Hopkins' colour theory is of great interest in its bearing upon the colour changes to which so much importance was attached by mediaeval alchemists in the operations of the Great Work, or preparation of the Philosopher's Stone (see Chapter III). 'According to the doctrine of Plato,' writes Hopkins,[28] 'the universals, such as matter, came first and upon them were impressed particular and individualising qualities. It was, therefore, deemed obligatory upon the alchemist, in his practical process of transmutation, as he wished for success, to begin with a material which was unidentifiable by particular qualities; and to impress upon it pure qualities which one after another should gradually rise in the scale of metallic virtue towards perfection.' This first material was often prepared by fusing together the four common base metals, lead, tin, copper, and iron. The resulting black surface-colour, characteristic of the first stage, was invested with such importance that it may have been the origin of the term Black Art, as applied to the operations of alchemy (*cf.* p. 4).

The next process, that of whitening, was accomplished by heating the black alloy with a little silver, followed by mercury (including arsenic and antimony) or tin. The third step was one of yellowing, for which the usual re-agents were gold, in small proportion, and sulphur water. In a final process, violet, the highest colour in this chro-matic hierarchy, was attained: Hopkins has shown how this colour, sometimes accompanied by a striking irid-escence, may be produced upon certain alloys containing small amounts of gold. Descriptions of operations leading to such sequences of colours permeate the writings of mediaeval alchemy. Usually, however, the sequences are more complex than the simple one just indicated, and the processes are veiled in mystical and allegorical language.

Another first material, having the requisite black hue, and attributed to Mary the Jewess, was prepared by fus-ing a lead-copper alloy with sulphur. This mysterious woman alchemist, to whom there are so many references in alchemical writings from the Alexandrian period on-wards, was often identified with Miriam, the sister of Moses. She has an excellent record of original work; for in addition to the discovery of Mary's Black, she is credited with the invention of the water-bath—still called *bain-marie* by the gallant French—and of the kerotakis.

The kerotakis, to which much importance is attached in the Greek writings, was a closed vessel in which thin leaves of copper and other metals could be exposed to the action of various vapours, including notably that of mercury; a continuous refluxing effect was secured by means of a condensing cover. In company with the alembic or still, ascribed to a certain Cleopatra,* it formed the chief laboratory apparatus of the early alchemists. The common alchemical references to ' things above and things below ', as in the Emerald Table of Hermes

* Not the queen; Cleopatra was an ancient Achaian name used by the Hel-lenising Macedonian dynasties, and meaning ' famously descended '.

(p. 55), are probably derived from the use of the kerotakis and other forms of circulatory stills.

In developing his colour theory, Hopkins [29] suggests that the two ' spirits ', mercury and sulphur, were the two substances most commonly used in the kerotakis, and herein he discerns the germ of the later mercury-sulphur theory of metals. Sulphur was at the same time the spirit of gold, by reason of its colour, and the spirit of fire, because of its combustibility. Further, gold, being convertible into a ' water ' by fusion, contained the spirit of liquidity. Thus, gold consisted of the opposed elements, fire and water, or, in a later sense, of the two ' spirits ', sulphur and mercury. The preparation of the brilliant red sulphide of mercury (known in Nature as cinnabar), by the union of sublimed sulphur and mercury in the kerotakis, therefore held a peculiar significance in the eyes of the Graeco-Egyptian alchemists. ' From this discovery ', observes Hopkins,[30] ' the sequence of colours begins to change in alchemistic literature from black, white, yellow, violet, to black, white (yellow), red.' ' Red is last in work of *Alkimy* ', wrote Norton in 1477 (p. 179).

Finally, according to this theory, the apotheosis of colour was reached in the Philosopher's Stone, ' the alchemistic form of Aristotle's entelechy, the final cause which can reproduce itself. For a complete analogy with Aristotle's philosophy, it was necessary that the alchemistic theory should conceive the Philosopher's Stone as the *grand finale* of the system.' [31]

If the main idea embodied in the colour theory be accepted, the germ from which alchemy developed must once more be sought in the early civilisations. Thus, according to Woolley, the famous Ziggurat, or ' high place ' of the gods, attached to the temple of Nannar the Moon-god at Ur of the Chaldees, and built in the twenty-third century B.C., was divided into zones of colour: these were black, red, blue, and golden, symbolising the division of the universe into the dark underworld, the habitable earth, the heavens, and the sun. Much later, the seven walls of Ecbatana, according to Herodotus, were painted in seven colours, arranged in the sequence, white (outermost), black, purple, blue, red, silvern, golden: these colours appear to have been associated with the seven heavenly bodies, and possibly also with the seven metals (p. 88).

§ 3. The Islamic Era

The Alexandrian era, during which the idea of the Philosopher's Stone appears to have been conceived, came to an end in the middle of the seventh century. The succeeding age, until the thirteenth century, witnessed a domination of alchemy by the Muslims. The writings of this period give evidence of notable advances, not only in alchemy, but also in mathematics, astronomy, and medicine. The practical achievements of alchemy, as described in the works of Geber (Jabir), Rhazes (al-Razi), Avicenna (Ibn Sina), and others, were enriched by the preparation of many new substances and by the skilful application of such processes as cupellation, reduction, distillation, calcination, and sublimation. Much of this experimental work, however, appears to have been based upon descriptions in the Graeco-Egyptian writings, and even in the still earlier works of Dioscorides.[32]

A very large number of the Islamic writings are ascribed to Geber, or Jabir,[33] a somewhat shadowy figure who appears to have flourished about the end of the ninth century. Jabir's chief practical aim was the preparation of the magic transmuting tincture, elixir, or powder of projection. The corresponding laboratory operations were correlated with the celebrated sulphur-mercury theory of the composition of metals which is expounded in these writings of Jabir, or of the Jabir school.

Aristotle had supposed that earth and water gave rise, respectively, to smoky and vaporous exhalations; the earthy smoke consisted of earth in process of change into fire, and the watery vapour was water undergoing conversion into air. Refractory stones and minerals consisted mainly of the earthy smoke, and fusible metals were formed from the watery vapour. Geber postulated an intermediate formation of sulphur and mercury from the exhalations, in the interior of the earth. Finally, sulphur and mercury, by combining in different proportions and in different degrees of purity, gave rise to the various metals and minerals (Fig. 2). The influence of

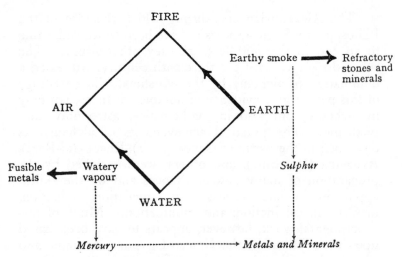

FIG. 2.—The Sulphur-Mercury Theory of the Origin of Metals.

(Geber's additions to the Aristotelian scheme are shown in italics.)

this view of the constitution of metals may be traced in the leading theories of chemistry until the overthrow of the theory of phlogiston by Lavoisier.

In considering possible relationships between the sulphur-mercury theory of the Muslim writers and the earlier Alexandrian school, Hopkins [34] states that the Egyptian theory regards the two 'tincturing spirits', mercury and sulphur, as endowed with a kinetic property which in operation led to transmutation. 'It does not visualise the conception of the metals as having a static quality composition (as by different combinations of mercury and sulphur) as one might mix two dye powders to produce a half shade such as purple. This latter is a different conception, a variation which came later and which we find expressed first probably by Geber in the ninth century, later certainly by the Pseudo-Avicenna and the Pseudo-Aristotle.' He adds that among the alchemists of Islam, who did not share the Egyptians' love of colour, ' the colour theory seems to have become very hazy, to have descended to a mere jargon of inherited terminology '.

Although the sulphur-mercury theory is perhaps based to some extent upon Graeco-Egyptian thought, its ultimate origin is obscure. In the final analysis it may perhaps be referred back to that primitive mode of thinking which is based upon a distinction of opposites, and it may be noted that even Aristotle's four ' qualities ' consisted of two pairs of opposites. Here, as so often in seeking alchemistic origins, we are forced to gaze back through indefinite vistas of time to the dim outlines of ancient civilisations. ' The doctrine of the two Contraries ', says Davis,[35] ' made its appearance in Chinese thought about the third century B.C., at the time when alchemy arose, and, like alchemy, it attached itself to Taoism. . . . The similarity of the Chinese doctrine to the earlier doctrines current in Mesopotamia and Egypt . . . suggests strongly that the Chinese doctrine derived from the earlier ones. But it does not at all indicate that Chinese alchemy likewise derived from earlier non-Chinese sources.'

In ancient India the Hindus held that the metals were born of the union of Hara (Shiva) and Parvati (the consort of Hara) through the help of Agni, the god of fire. Mercury was associated with the semen of Hara, sulphur with Agni, and earth (or crucible) with Parvati.[36] A similar idea takes a prominent place also among the philosophical concepts of ancient China. In an interesting introduction to the English version of the *Ts'an*

PRELUDE TO CHEMISTRY

T'ung Ch'i of Wei Po-yang (p. 7), Davis [35] points out that Chinese alchemy is based upon the two fundamental ideas of the Five Elements (*Wu-hsing*) and the Two Contraries (*Yin-Yang*). The five elements—which were recognised as early as the twelfth century B.C.—are water, fire, wood, gold (or metal), and earth; these were correlated with the five virtues, tastes, colours, tones, seasons, and numerous other magic quintets.

The original conception of *Yin-Yang* dates back to about the sixth century B.C. In the later Taoist interpretation, the so-called *t'ai-chi*, or first matter of the universe, eventually gave rise to two principles possessing opposite characters: *Yin* was feminine, negative, heavy, and earthy; while *Yang* was masculine, positive, light, and fiery. *Yang* donates and *Yin* receives, says Wei Po-yang; [37] while, according to Ch'ang-ch'un,[38] *Yang* and the sky are masculine and their element is fire, whereas *Yin* and the earth are feminine and their element is water. Ch'ang-ch'un adds that *Yin*, the imperfect, can quench *Yang*, the perfect: wherefore it behoves the Taoist to lead a careful life. By their interaction, the two contraries were held to give rise to the five elements constituting the material of the world. *Yang* and *Yin* were associated with the Sun and Moon, respectively; and the five elements, in the order given above, with the planets Mercury, Mars, Jupiter, Venus, and Saturn.

As indicated above (p. 19), this doctrine of the two contraries, or reciprocal principles of Nature, may have come to China from Babylonia or Egypt.[39] The two principles—usually depicted as active and passive, or male and female, and often symbolised as Sun and Moon—were an expression of the fundamental natural law of reproduction or procreation. They find a place, under different names, in the mythological systems of all the ancient civilisations; moreover, in various mystic unions, such as that of soul and body, they were venerated and worshipped as hermaphroditic deities. Thus, in the mythology of ancient Egypt, the Sun-god, Osiris, was regarded as the active principle, vivifying influence, or originating source of energy and power; Isis, the Moon-goddess, was the passive recipient and producer, symbolising terrene nature: Horus, the annual offspring of these two deities, represented the infant year and season of growth.

From what has been said, the essential identity of the ancient Chinese doctrine of *Yin-Yang* with the sulphur-mercury theory of the Western civilisations is evident. Sulphur, like *Yang*, was linked with the Sun; and mercury, like *Yin*, with the Moon. Also, like Osiris and Isis, the Sun-god and Moon-goddess of ancient Egypt, these principles of the Sun and Moon were held to give rise to all things by their conjunction or interaction.

The Sun-god and Moon-goddess; Yang and Yin; masculine and feminine; sulphur and mercury; positive and negative; proton and electron: truly, it may be said of chemical theory that the more it changes the more it is the same thing! The Doctrine of the Two Contraries seems to make a peculiar appeal to some deep-seated instinct in the human mind. As Hoefer remarks in more general terms, the systems which confront the intelligence remain basically unchanged through the ages, although they assume different forms; thus, through mistaking form for basis, one conceives an unfavourable opinion of the sequence. We must remember, he adds, that there is nothing so disastrous in science as the arrogant dogmatism which despises the past and admires nothing but the present.*

§ 4. The European Era

From the twelfth century onwards, the alchemical literature of the Islamic school began to percolate by way

* ' Ce sont au fond toujours les mêmes systèmes qui, à différentes époques, se présentent à l'esprit, revêtus seulement de formes différentes; puis, confondant la forme avec le fond, on porte sur le tout un jugement défavorable . . . nous devons rappeler qu'il n'y a rien de plus funeste à la science que l'orgueil du dogmatisme qui dédaigne le passé et n'admire que le présent ' (F. Hoefer, *Histoire de la Chimie*, Paris, 1866, i, 249).

of Spain into western Europe, through the medium of Latin translations of the Arabic texts (p. 42). The new learning was eagerly embraced by the schoolmen, and during the next few centuries numerous European writings led to a wide dissemination of alchemical ideas in this fresh field. At the same time, however, the unfamiliar nature of alchemy and its strange jargon of terms led to much confusion of thought in mediaeval Europe. The ideas and allusions of the original writers were often distorted into unintelligible shapes by copyists and commentators who were satisfied to accept literal interpretations of cryptic statements and to play with an impressive but meaningless alchemical vocabulary.

One idea, however, was transmitted with an all too literal accuracy (p. 13): this was the imagined possibility of producing gold artificially. The seed of this idea, germinating in the credulous mediaeval mind, expanded into an arborescent growth beneath whose shade ' the armies of powerful Sultans might repose '. In the heyday of gold-making, during the sixteenth century, Paris, Prague, and other European cities contained numerous alchemical workshops and dens in which this strange craft was prosecuted with feverish vigour. The ancient houses of the gold-makers are still to be seen at Prague in the Zlatá Ulička, or Golden Alley—the street frequented by alchemists of the time of the Emperors Maximilian II (1564–1576) and Rudolph II (1576–1612). The possibilities and malpractices of gold-making were viewed so seriously by those in authority that during the fifteenth century alchemistic activities were interdicted in various countries, including England, and the making of gold and silver was prohibited by law. Mercenary minded gold-seekers, fastening avidly upon the allurements of multiplying gold, degenerated into quacks and charlatans, and eventually brought alchemy into disrepute through their false claims and misguided activities. ' For ever we lacken oure conclusioún ', laments the disillusioned Canon's Yeoman, and continues: [40]

To moche folk we bring but illusioún,
And borrow gold, be it a pound or tuo,
Or ten or twelve, or many somm*e*s mo,
And make them thinken at the least*e* weye,
That of a single pound we can make tweye.
Yet is it fals; and ay we have good hope
It for to do, and after it we grope.
But that sciénce is so far us bifore,
We never can, although we had it sworn,
It overtake, it slideth away so fast;
It wol us mak*e* beggers att*e* last.

It would be a mistake, however, to regard Chaucer's
'canoun' and his fellow 'pseudo-alchemists' as the
only representatives of mediaeval alchemy. More credit
attaches to the thought and work of contemporaneous
alchemists of the philosophical and mystical types. By
some writers on alchemy these are depicted indeed as the
'true adepts', to whom practical alchemy was a branch
of a comprehensive philosophical system. According to
this view, the experimental attempts of the adepts to transmute metals were carried out with the aim of adducing a
material proof of their system: on the one hand were
ranged the sheep, the elect fraternity of alchemical cognoscenti; on the other stood the goats, the worldly-minded
seekers after gold, those alchemical 'outsiders' sometimes disdainfully known as 'souffleurs', or 'puffers'.
That there were various types of alchemists is clear; but
sometimes the distinctions obtaining between them have
been defined too rigidly. Even as late as 1850, for
example, the true adepts, the favoured sheep, were
characterised by the writer of *A Suggestive Inquiry into
the Hermetic Mystery* [41] as 'truthful, moral, always pious
and intelligent', and actuated by motives 'of the purest
possible kind'; while the rejected goats, the despised
'pseudo-alchemists, on the other hand, were reckless and
despicable'.

All in all, whether of the sheep or of the goats,
mediaeval alchemy may fairly be summed up as 'much
cry and little wool'. Accident brought to it certain

23

spectacular discoveries, such as that of gunpowder; but beyond this, to quote Hopkins,[42] 'with the exception of the recognition of the mineral acids, to which may be added certain empirical facts developed in the mining industry, the experimental knowledge of the thirteenth and fourteenth centuries was little in advance of that of the third century of our era'. To this statement may be added the opinion that alchemical theory deteriorated rather than advanced during the mediaeval era.

Mediaeval works on alchemy abound in references to the sulphur-mercury theory. Thus, in the thirteenth century, the pseudo-Roger Bacon wrote as follows in his *Speculum Alchemiae*:[43] '*Alchimy* is a Science, teaching how to transforme any kind of mettall into another: and that by a proper medicine, as it appeareth by many Philosophers Bookes. *Alchimy* therefore is a science teaching how to make and compound a certaine medicine, which is called *Elixir*, the which when it is cast upon mettalls or imperfect bodies, doth fully perfect them in the verie projection. . . . The natural principles in the mynes, are *Argent-vive*, and *Sulphur*. All mettalls and minerals, whereof there be sundrie and divers kinds, are begotten of these two: but I must tel you, that nature alwaies intendeth and striveth to the perfection of Gold: but many accidents coming between, change the mettalls. . . . For according to the puritie and impuritie of the two aforesaid principles, *Argent-vive*, and *Sulphur*, pure and impure mettalls are ingendred.'

A representation of Thomas Aquinas (Plate 3), appearing in a work [44] published in 1624, forms an apt illustration of the above description of thirteenth-century alchemy. It contains a pictorial representation of the natural production of metals in the bowels of the earth, from exhalations marked with the signs of sulphur and mercury; above, on the crust of the earth, an alchemist is engaged in an experimental attempt to make the same metals artificially. The accompanying Latin epigram

PLATE 3

Thomas Aquinas the Italian.

From *Viridarium Chymicum*, Stolcius, 1624 (reprinted from *Symbola Aureae Mensae Duodecim Nationum*, Maier, 1617). (See pp. 24, 223.)

PLATE 4

Mare eff Corpus, duo Pifces funt
Spiritus & Anima,.

Body, Soul, and Spirit.

From the *Book of Lambspring,* in *Musaeum Hermeticum,* 1625. (See p. 28.)

PLATE 5

The Fair Flower on the Mountain, with the Dragons and Griffins of the
North.

From the Figures of Abraham in the MS. *Philosophorum Praeclara Monita.*
(See pp. 47, 62, 63.)

PLATE 6

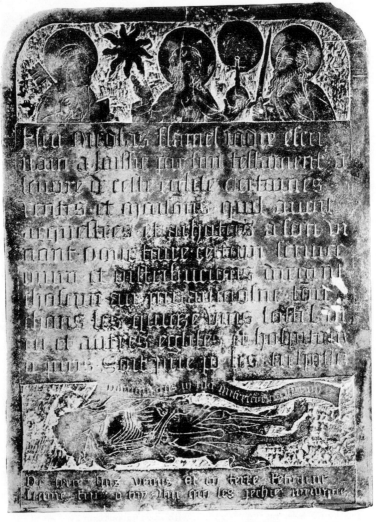

An inscribed marble tablet from the tomb of Nicolas Flamel.
Early 15th century.   In the Musée de Cluny, Paris.

states that ' from mercury mingled with its own sulphur, art, even as Nature, brings forth all metals '.

The mercury and sulphur of this theory were not held to be identical with the common substances bearing these names. It was already known to Geber that under experimental conditions these two substances gave rise to the ' red stone ' known in nature as cinnabar (*i.e.* mercuric sulphide), and not to metals. The names stood rather for combinations of properties or qualities: for example, sulphur was sometimes held to typify visible properties, such as colour, while mercury represented invisible or occult properties, such as fusibility. The view was widely held among mediaeval alchemists that when ordinary mercury (or silver) and gold were divested of their grosser (physical) properties, they yielded ' philosophical ' or ' sophic ' mercury and sulphur—that is to say, the mercury and sulphur of the theory. Sophic mercury was often called ' fixed ' mercury, that is, the solid essence or core of mercury—corresponding, perhaps, in a more modern parlance, to a mercury whose atom has been divested of certain electrons. Aristotle's theory provided four fundamental causes of the physical properties of substances, and it may be said that broadly the sulphur-mercury theory supplemented the older theory by providing two fundamental causes of chemical properties. Essentially, mercury stood for metallicity and sulphur for inflammability. In philosophical sulphur is to be discerned, indeed, the germ of the phlogiston theory of the seventeenth and eighteenth centuries.

A revealing summary of the sulphur-mercury theory, as it was interpreted at the beginning of the seventeenth century, is given by Ben Jonson's alchemist, Subtle.[45] He explains to Mammon that the ' elementary matter of gold ', and of all metals and stones, is formed by incorporating a humid exhalation, or unctuous water, with a viscous earth. He then continues, in terms which bear a strong resemblance to the pseudo-Roger Bacon's account of some three hundred years earlier:

Where it is forsaken of that moisture
And hath more dryness, it becomes a stone.
Where it retains more of the humid fatness,
It turns to sulphur, or to quicksilver,
Who are the parents of all other metals.
Nor can this rémote matter suddenly
Progress so from extreme unto extreme,
As to grow gold, and leap o'er all the means.
Nature doth first beget th' imperfect, then
Proceeds she to the perfect. Of that airy
And oily water mercury is engendered;
Sulphur o' the fat and earthy part.

Alchemical theory was extraordinarily static in its essentials.   Naturally, however, throughout the centuries during which it held sway, the sulphur-mercury theory was interpreted in manifold ways.   A slight acquaintance with alchemical literature is sufficient to furnish some idea of the labyrinthine theoretical and practical schemes which the writers constructed by associating this theory with the conception of the Philosopher's Stone and with many other ideas derived from the doctrines and beliefs of astrology, mystical theology, mythology, magic, and the like.   Thus, the ' mercury of the philosophers '—the essence or soul of the metals, divested of the gross physical properties represented by earth, air, fire, and water—was often identified with the *prima materia* of the Ionian philosophers.   Strip a metal of these qualities, and it yields the one primitive matter; impose the appropriate new qualities upon the primitive matter, and the desired new substance is attained.   ' The metals are similar in their essence, and differ only in their form ', wrote Albertus Magnus in *De Alchimia;* and in another work he added, ' one may pass easily from one to another, following a circle '.   The modern electronic theory of matter expresses a similar conclusion.

It was probably the alchemical doctrine of the unity of all things which led to a recognition of mercury and sulphur as the counterparts of man's spirit and soul. Later, the original sulphur-mercury theory was extended by the addition of salt (or ' magnesia ') as the third member of the so-called *tria prima*, or three ' hypostatical principles '. This system is commonly represented by a triangle in alchemical symbolism. Salt represented materially the principle of uninflammability and fixidity, and mystically the body of man.

A definite association of mercury, sulphur, and salt with spirit, soul, and body respectively was made by Paracelsus [46] in the following words: ' Know, then, that all the seven metals are born from a threefold matter, namely, Mercury, Sulphur, and Salt, but with distinct and peculiar colourings. . . . Mercury is the spirit, Sulphur is the soul, and Salt is the body . . . the soul, which indeed is Sulphur . . . unites those two contraries, the body and the spirit, and changes them into one essence.' The harmonies [47] of the *tria prima* may be summarised as follows:

Mercury	*Sulphur*	*Salt*
Metallicity, fusibility, volatility	Inflammability	Uninflammability and fixidity
Volatile, and unchanged in the fire	Volatile, and changed in the fire	Found in the ashes
Spirit	Soul	Body
Water	Air	Earth

These relationships, presented in many forms, are among the most familiar features of alchemical literature. According to Paracelsus,[48] who first gave a tangible form to the theory of the *tria prima*, ' The three principles . . . from which all things are born and generated ' are ' phlegma, fat, and ash. The phlegma is Mercurius, the fat is Sulphur, and the ash is Salt. For that which smokes and evaporates over the fire [*e.g.* in the burning of wood] is Mercury; what flames and is burnt is Sulphur; and all ash is Salt.' Basil Valentine [49] gives a similar example: ' If a rectified *Aqua vitæ* be lighted, then Mercury and the vegetable Sulphur separateth, that Sulphur burns bright, being a meer fire, the tender Mercury betakes himself to his wings and flieth to his *Chaos* '.

As an example of another type, the quaint *Book of Lambspring* (p. 167) contains an illustration (Plate 4) of two fishes swimming in the sea: ' the Sea is the Body, the two Fishes are Soul and Spirit '. Another illustration in this book depicts a deer and a unicorn in a forest, and the accompanying interpretation states that ' in the Body [the forest] there is Soul [the deer] and Spirit [the unicorn] . . . he that knows how to tame and master them by Art, to couple them together, and to lead them in and out of the forest, may justly be called a Master, for we rightly judge that he has attained the golden flesh '.

Poisson [50] gives the following synthetic view of alchemical theory:

Primary Matter unique, indestructible	Sulphur fixed principle	Earth (visible, solid state)
		Fire (occult, subtle state)
	Salt	Quintessence (state comparable to the ether of physicists)
	Mercury volatile principle	Water (visible, liquid state)
		Air (occult, gaseous state)

Salt is here regarded merely as a medium uniting sulphur and mercury, in a manner similar to the supposed union of body and soul by means of the vital spirit; often salt was dismissed as a supernumerary principle.

Pseudo-alchemy reached its zenith between the fourteenth and sixteenth centuries. The great growth of alchemical literature which took place in this period has been characterised by Lacroix [51] as a chaotic collection of absurdities, in which ' everything grand or mysterious was attributed by the alchemists to the demons which people the air, fire and water, to the stars which are superior to the human and to the Divine will, to mysterious sympathies existing between the Creator and his creatures, and to the hybrid combinations of mineral and vegetable substances '.

Thus alchemy, as a whole, developed into what

AN OUTLINE OF ALCHEMY

Poisson [52] has termed the most nebulous scientific heritage of the Middle Ages: ' Scholasticism with its infinitely subtle argumentation, Theology with its ambiguous phraseology, Astrology so vast and so complicated, are only child's play in comparison with Alchemy '.

At the beginning of the sixteenth century, alchemy therefore presents a complicated pattern which may be deciphered only incompletely and with difficulty. Its theory consisted of a chaotic mass of ideas derived from every conceivable source and projected upon a confused background of late mediaeval adaptations of the writings of Aristotle and Pythagoras. Its practitioners were occupied chiefly with indiscriminate searchings after wonderful elixirs and magic transmuting agents, and the operations of alchemy had become increasingly associated with charlatanism and fraud. At the same time, as may be seen from an examination of such works as the *Sum of Perfection* (p. 47), the *Codex Germanicus* (p. 75), Brunschwick's *Book of Distillation* (p. 75), and Georgius Agricola's *De Re Metallica* (p. 77) of a somewhat later date, another type of alchemist was soberly concerned in the application of alchemical processes and knowledge in mining, metallurgy, the isolation of the essential oils of plants, and other useful arts. Such was alchemy as Paracelsus found it; and he was instrumental in subjecting it to a fresh tendency.

Paracelsus (1493–1541), besides developing the theory of the *tria prima*, introduced a new era of iatrochemistry, or alchemy in the service of medicine. ' Marvellous Paracelsus, always drunk and always lucid, like the heroes of Rabelais ', expounded with great vigour, seasoned with venom, the view that the chief object of the alchemist should be to aid the apothecary and the physician, rather than to make gold. His great achievement for alchemy lay not in the introduction of revolutionary theoretical ideas or practical advances, but in his determined attempt to liberate the incipient science from the narrow and sordid domination of the ' multipliers '

29

and ' bellows-blowers '; he gave alchemy a new orientation and endowed it with a fresh stimulus. Paracelsus was essentially a propagandist of a new order, both in alchemy and medicine; as such he aroused vigorous opposition, especially from the physicians, whom he loaded with abuse so violent and coarse that the Scottish historian of chemistry, Thomas Thomson, after reaching a certain point, found himself unable to proceed with its translation.[53] In 1520, Paracelsus' contemporary reformer, Luther, symbolised the freeing of religion from the trammels of orthodoxy and tyrannic authority by burning the papal bull at Wittenberg; five years later, Paracelsus' public burning of the works of Galen and Avicenna before the burghers and physicians of Basel constituted a similar gesture in the realms of medicine and alchemy.

Liberated gradually from its main obsession of gold-making, alchemy changed imperceptibly, during the next few generations, into chemistry. Lacroix [54] expresses the change in the following words: ' Owing to the labours of Paracelsus, alchemy exchanged its speculative for a practical character; and . . . George Agricola . . . effected, without any disturbance or noisy discussion, the auspicious revolution in metallurgy which his ardent contemporary was unable to achieve without a fierce struggle in medicine. . . . Henceforth, *chimiastrie* [iatrochemistry], or the art of transforming bodies and substances from a medical point of view, and metallurgy, or the art of extracting and purifying metals for the use of industry—two sciences having many points of contact and of difference—advanced in parallel lines upon the road of progress. Alchemy, ceasing to be experimental and becoming merely psychological, was abandoned to the study of a few fanatical adherents, and finally disappeared altogether from the enlarged domain of positive science.' ' The science of chemistry ', says Hoover,[55] ' comes from three sources—alchemy, medicine and metallurgy '.

The work of Libavius (1540–1616), van Helmont

(1577–1644), and Glauber (1604–1668) furnished numerous valuable additions to the practice of chemistry. In 1595 Libavius published his celebrated *Alchymia*, which served as an authoritative and comprehensive textbook to his immediate successors; van Helmont was the first to recognise gases as distinct substances; and Glauber added much to the existing knowledge of metals, acids, and salts. These chemists, like their great successor, Robert Boyle, shared the alchemical belief in transmutation, but they no longer regarded gold-making as their main goal. Alchemical mysticism and obscurantism were gradually fading away, and although alchemical traditions lingered until late in the eighteenth century, the long age of alchemy (Fig. 3) virtually ended with Robert Boyle's abolition of the systems of the four elements and the three hypostatical principles, and his introduction of the modern chemical idea of an element, as expounded in his celebrated book published in London, in 1661, under the title of *The Sceptical Chymist*.[56]

§ 5. THE ERA OF PHLOGISTON

Modern chemistry did not spring fully equipped from between the covers of *The Sceptical Chymist*, as there is sometimes a tendency to imply. It is true that the outer defences of the alchemical citadel were breached by Boyle, but some four generations elapsed before its main works were carried by Lavoisier: Boyle's *Sceptical Chymist* appeared in 1661, Lavoisier's *Traité Élémentaire de Chimie* in 1789.

'The Experiments wont to be brought, whether by the common Peripateticks, or by the vulgar Chymists, to demonstrate that all mixt bodies are made up precisely either of the four Elements, or the three Hypostatical Principles, do not evince what they are alledg'd to prove . . . I now mean by Elements, as those Chymists that speak plainest do by their Principles, certain Primitive and simple, or perfectly unmingled bodies; which not being made of any other bodies, or of one another, are the Ingredients of which all those call'd perfectly mixt Bodies are immediately compounded, and into which they are ultimately resolved.' [57]   So run Boyle's words.

Boyle swept away the ancient Graeco-Egyptian theoretical system with its hoary concretions, and advanced an essentially modern definition of an element.   Nevertheless, for another century or so, Boyle's ideas were neglected and met with no practical application; the doctrine of transmutation lingered on, and the fundamental nature of air, water and fire remained unknown. Although formally banished from the stage, the four Aristotelian ' elements ' and the three hypostatical principles still dominated from behind the scenes that period of transition from alchemy to chemistry which is called the era of phlogiston.

Two of these ' elements ', fire and air, had long been associated with the phenomenon of burning, or combustion.   This, of all familiar chemical processes, is the most spectacular and impressive; yet it was not until the second half of the seventeenth century that any serious attempt was made to find a chemical explanation for the phenomenon.   That explanation, conceived in the form of the theory of phlogiston, is clearly alchemical in nature and derivation: the phlogiston theory may indeed be called the swan song of alchemy.[58]   It is unnecessary for the present purpose to detail the rise of the phlogiston theory at the beginning of the eighteenth century and its overthrow by Lavoisier towards the end of the same century,[59] but it is of interest to consider briefly the relationship of this outstanding theory of the transition period to the preceding age of alchemy.

Becher (1669) held that the ' calces ' which result when metals like lead and tin are heated in air are formed through the expulsion from the metal of a combustible principle, or fatty earth, which he called *terra pinguis*. His pupil Stahl (1702) associated combustibility with the presence in the combustible body of a constituent which he called *phlogiston*. Burning was thus equivalent to a loss of phlogiston, and a metal was regarded as a compound of its calx with phlogiston.

To the student of alchemy it is clear that phlogiston has an ancestry as old as alchemy itself. ' The Sun and Moon of Hermes ', writes Davis [60] in an illuminating passage, ' are the same as the positive Sulphur and negative Mercury principles of Zosimos and Jabir. The Latin Geber said that sulphur is *pinguedo terrae* (fatness of the earth). Becher thought that the Sulphur principle ought to be called *terra pinguis* (fat earth), and his disciple, Stahl, preferred to name it *phlogiston*. The doctrine of the two contraries, with only superficial modifications, dominated the science of chemistry until the time of Lavoisier.'

How many chemists realise that phlogiston was a direct descendant of Osiris (p. 21), the Sun-god of ancient Egypt!

Whatever the demerits of the phlogiston theory, it had the merit of co-ordinating many observations which had hitherto been viewed in isolation; moreover, it was a definite and comprehensible theory which stimulated enquiry on scientific lines. Under its aegis, a remarkable band of eighteenth-century investigators accumulated the evidence which led to its fall. Thus, Black first weighed a gas (' fixed air ' or carbon dioxide) in combination. Priestley, on August 1st, 1774, perhaps the most significant date in the history of chemistry, isolated ' dephlogisticated air ' (oxygen), that ' fire-air ' of Scheele which may be regarded as the antithesis of phlogiston; moreover, with the aid of his pneumatic trough, he obtained many other ' Different Kinds of Air '.[61] Cavendish, working with a quantitative precision hitherto unattained, ' weighed the Earth, analysed the Air, discovered the compound nature of Water, and noted with numerical precision the obscure actions of the ancient element, Fire '.[62] The cool, unbiassed mind of Lavoisier, with its keen temper of French logic, was quick to perceive the

implications of these wonderful experimental results.
The discovery of the composition of the ancient 'elements', Air and Water, sounded the last note in the
swan song of alchemy. The lingering vestiges of the
Greek system vanished with phlogiston into thin air, 'and
on the mere the wailing died away'. The new age had
come.

Hermes Trismegistos. From a temple
at Pselcis (*c.* 250 B.C.). The symbolism
is noteworthy. The figure holds the
'*ankh* or *crux ansata* in one hand, and
in the other a staff with a serpent,
scorpion, hawk's head, and asp enclosing a circle.

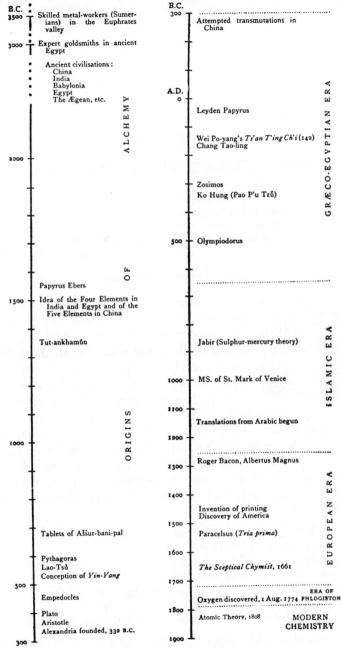

FIG. 3.—Prelude to Chemistry: a Time Chart.

35

CHAPTER II

THE LITERATURE OF ALCHEMY

I'll show you a book, where Moses and his sister,
And Solomon have written of the art;
Aye, and a treatise penned by Adam . . .
O' the philosopher's Stone, and in High Dutch.

BEN JONSON

§ 1. GENERAL REVIEW

 OTHING is more characteristic of alchemy than its literature. Indeed, many of the finer aspects of 'the Divine Art' become evident only to the wanderer through the wide domain of the alchemical writers. The rest of this book may serve to show that such a wanderer, even though he travel at haphazard, will not lack entertainment. A veritable 'pleasure garden of chemistry' will open its gates to the 'wearied Servants of Laboratories' who seek their recreation within this domain: a garden in which the decorated manuscripts and fire-stained tomes of a long-forgotten age will put on the garments of youth and flower afresh.

But if thou wilt enter this *Campe of Philosophy*
With thee take *Tyme* to guide thee in the way;
For By-pathes and Broad wayes deepe Valies and hills high
Here shalt thou finde, with sights pleasant and gay,
Some thou shalt meete with, which unto thee shall say,
  Recipe this, and that; with a thousand things more,
  To *Decipe* thy selfe, and others; as they have done before.[1]

Probably the oldest writing extant of alchemical interest is the Papyrus Ebers, which was found in a tomb

36

at Thebes about 1862. This 'oldest book in the world' is now preserved in the University of Leipzig. It consisted, when discovered, of a roll of papyrus twelve inches wide and sixty-eight feet long. It is written in Hieratic, the script of the priests, which is a cursive form of the chiselled hieroglyphics of ancient Egypt. Through the earlier labours of Thomas Young and Champollion on the hieroglyphical rendering of the related Demotic script of the Rosetta Stone, it became possible for later investigators to work the meaning of the Papyrus Ebers 'out of the inky darkness in which it had remained for thousands of years'.[2] It dates from about 1550 B.C., and portions of its contents may be some twenty centuries older. It is mainly a collection of prescriptions, but Ebers mistakenly claimed it as one of the 'Hermetic Books' of ancient Egypt, the authorship of which was ascribed to Hermes or Thoth.

Altogether, the Papyrus contains 811 prescriptions, most of which give rise to interest by their novelty rather than to confidence by their nature. For example, if the hair is turning grey, it should be anointed with a black calf's blood, boiled in oil; if it is falling out, it should be treated with the mixed fats of the horse, hippopotamus, crocodile, cat, snake, and ibex; if it is weak, it should be strengthened through the application of a crushed donkey's-tooth, disseminated in honey. Again, the powdered tooth of a hog, put inside four sugar-cakes, is a remedy for indigestion: the magic number 4, which was adopted later by the Pythagoreans, occurs repeatedly in the Papyrus. Among the mineral ingredients mentioned are stibnite, calamine, granite, sulphur, soda, lead, copper, verdigris, lapis lazuli, salt, and saltpetre. The appeal of many of these ancient 'remedies' was evidently connected with their revolting nature; some of them are unquotable.

There are also various Babylonian tablets extant, which deal with such matters of chemical interest as the formulae used by workers in glass and metals.[3] The royal library of Aššur-bani-pal, king of Assyria in the seventh century B.C., contained a series of such tablets; these indicate that the Assyrians had an extensive empirical knowledge of chemical operations. The vocabulary of modern chemistry still contains a light sprinkling of names derived from this ancient civilisation: among them are ' alcohol ' (from *guhlu*, eye-paint), ' naphtha ' (from *naptu*), and ' sandarach ' (from *šindu arqu*).

In China the earliest known treatise of a purely alchemical character is the *Ts'an T'ing Ch'i*, written by Wei Po-yang about 142 A.D.: this deals with the preparation of a medicine capable of making living beings immortal (p. 121). Its contents show that an established alchemical tradition existed in China in the second century A.D. There are anterior Chinese writings containing references to alchemy; but the *Ts'an T'ing Ch'i* is stated to be the earliest written work in Chinese, or any other language, which is concerned solely with alchemy. It was translated [4] into English in 1932. Wei Po-yang (Plate 20), who has been called ' the father of Chinese alchemy ', was a Taoist philosopher and alchemist; he described himself as ' a lowly man from the country of *Kuai*, who has no love for worldly power, glory, fame, or gains, who wastes his days leading a simple, quiet, leisurely, and peaceful life in a retreat in an unfrequented valley '. Definite points of similarity may be traced between the *Ts'an T'ing Ch'i* and the writings of Western alchemy (pp. 7, 122); the same may be said of the later Taoist works on alchemy, such as those of Ko Hung (p. 6).

Apart from the general chemical interest of such literary remains of the ancient civilisations as the Papyrus Ebers and the Assyrian tablets of Aššur-bani-pal, the written records of occidental alchemy begin with Greek texts derived from the Alexandrian era. Of surviving manuscripts [5] written in Greek, the oldest of any note are the so-called Leyden and Stockholm papyri, which were written in Egypt in the early centuries of the Christian era. The Leyden papyrus, dating probably

THE LITERATURE OF ALCHEMY

from the first or second century A.D., is a collection of a
large number of recipes dealing mainly with the preparation of metals and alloys simulating gold and silver and
with processes for augmenting these precious metals.
The Leyden papyrus also contains the earliest account of
the preparation of ' water of sulphur ' (p. 14). Most
of the contemporary papyri are said to have been destroyed
by the Romans, following a decree of the Emperor Dio-
cletian abolishing alchemy from Egypt (292 A.D.).

The remaining manuscripts deriving from the Graeco-
Egyptian era are transcripts of earlier documents. The
oldest of them is the manuscript of St. Mark of Venice,
which does not date back farther than the tenth or eleventh
century. Others are as recent as the sixteenth and seventeenth centuries. These manuscripts are preserved in the
Bibliothèque Nationale at Paris, and other European
libraries. They were first translated and studied by and
for Marcellin Berthelot, the results of whose work are
contained in *Les Origines de l'alchimie* (1885), *Collection
des anciens alchimistes grecs* (1888), *La Chimie au moyen
âge* (1893), and other publications. The Greek originals
of these manuscripts appear to have been composed in
Egypt between the first and seventh centuries A.D. Some
of the compositions are the work of authentic authors,
but others bear ascriptions to Hermes, Isis, Democritus,
Moses, Mary the Jewess, and other exalted and sup-
posedly influential personages: such attributions, like the
fulsome dedications of a later age, were probably made in
order to secure prestige for the work.

Some fifteen hundred years later, Ben Jonson made a
shrewd hit at this type of alchemical literature in *The
Alchemist*,[6] when he caused Mammon to exclaim:

Will you believe antiquity? records?
I'll show you a book, where Moses and his sister,
And Solomon have written of the art;
Aye, and a treatise penned by Adam . . .
O' the philosopher's Stone, and in High Dutch . . .
Which proves it was the primitive tongue.

39

PRELUDE TO CHEMISTRY

The chief writers of the early Greek manuscripts are the pseudo-Democritus (1st or 2nd century), Synesius (2nd or 3rd century), Zosimos (3rd or 4th century), Olympiodorus (5th or 6th century), and Stephanus (7th century).   Their contributions to the theory and practice of alchemy have been summarised in Chapter I.   The pseudo-Democritus is interested primarily in the production of imitative gold and silver, for which he gives numerous recipes; he indicates that some of these methods were used by the priests of ancient Egypt.   Both this writer and Synesius comprehend such processes under the name of 'the Divine Art', thereby suggesting a close connection between alchemy and religion in ancient Egypt.

The most important of these authors is Zosimos of

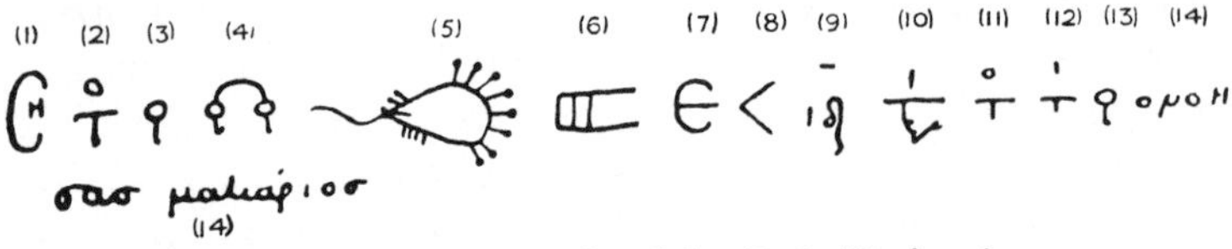

FIG. 4.—The Formula of the Crab (Zosimos).

Panopolis, who evinces a philosophic interest in the current technical processes.   He also refers to Cleopatra, Mary the Jewess, and other writers, real or imaginary, and describes the still and the kerotakis.   Various drawings of such early forms of alchemical apparatus occur in the margins of the manuscript of St. Mark of Venice, and some of these are reproduced in the works of Berthelot. The same manuscript bears a later annotation purporting to interpret the celebrated Formula of the Crab, which is given in the writings of Zosimos (Fig. 4).   This Formula, which was reported to embody the secret of transmutation, was probably a cipher used by Egyptian craftsmen engaged in making imitative gold (p. 154).   It is of particular interest as a precursor of alchemical symbolism.   The first symbol is said to be equivalent to 'message begins'.   The second is a contraction of $\tau\grave{o}$

40

PLATE 7

The Figures of Abraham the Jew.

From *Nicolas Flamel*, Poisson, 1893.

PLATE 8

Testing the Purity of Sulphur in the 14th century.

From the *Codex Germanicus, c.* 1350.   (See p. 75.)

PLATE 9

Salt Manufacture in the 16th century.

From *De Re Metallica*, Agricola, 1561.

A—shed.
B—painted signs.
C—first room.
D—middle room.

E—third room.
F—two little windows in the
    end wall.
G—third little window in the
    roof.
H—well.

I—well of another kind.
K—cask.
L—pole.
M—forked sticks in which the
    porters rest the pole
    when they are tired.

PLATE 10

Was helffen Fackeln / Liecht oder
Brillen /
Wann die Leut nicht sehen wollen.

A Device of Heinrich Khunrath.

From his *Von Hylealischen, das ist Primaterialischen
Catholischen oder Algemeinem Natürlichen Chaos*,
Magdeburgk, 1616. (See p. 81.)

πᾶν, that is ' the all ', or the universal matter: here it refers to an alloy of lead and copper, constituting the first material of Mary the Jewess (p. 15).   The third symbol means verdigris (' rust of copper '); the crab is the symbol of fixation; a reference to the philosopher's egg (p. 149) has been extracted from the tenth symbol; and the last one, appropriately enough, has been interpreted as meaning ' blessed is he who gets understanding '.[7]

When the manuscripts are considered in chronological order, the practical aspect of alchemy appears to become progressively more remote.   According to Holmyard,[8] the writings ' offer us the most bizarre picture of Gnostic theory intermingled with chemical fact, ecstatic visions, descriptions of apparatus, and injunctions to the reader to keep the secret of the Art from the vulgar. . . . As we pass to those chemists who succeeded Zosimos—Pelagius, Synesius, Heliodorus, Olympiodorus, and others—speculation and occult theory grow ever wider apart from experimental fact, and at length we encounter the conception of a philosopher's stone.'

Many alchemical manuscripts written in Arabic and Syriac, during the eighth century and later, are preserved in the great libraries of the world, and occasional discoveries of uncatalogued manuscripts of this kind are still being made in Egypt and the adjacent countries. These came into being in the following way.   With the close of the Alexandrian era (c. 650) and the rise of Islam, copies and translations of many Greek and Graeco-Egyptian writings on alchemy, mathematics, medicine, philosophy, etc., were made in Persia and Syria.   Between the eighth and the thirteenth centuries great schools of learning arose at Baghdad, Damascus, Cordova, Toledo, and other centres of Muslim culture.   In due course, translations from the Greek were supplemented by the indigenous writings of such Muslim savants as Geber (' the greatest chemist of Islam '), Rhazes (' the Persian Boyle '), and Avicenna (' the Aristotle of the Arabians '). The general character of these writings has been outlined above (p. 17). Through their eventual translation into Latin, much of the academical knowledge of the Arabs, the Greeks, and the ancient civilisations was transmitted to western Europe. Geber's *Book of Seventy* and *Book of Mercy* are known in both versions, Arabic and Latin; but some of the most important of the Latin works attributed to him, including the celebrated *Sum of Perfection* (p. 47), have no known Arabic originals, so that their authenticity is not proven.

European scholars first became interested in the translation of alchemical texts early in the twelfth century; such translators as Gerard of Cremona, Michael Scot, Robert of Chester, Adelard of Bath, and Hermann of Carinthia were thus instrumental in making known to western Europe the accumulated wisdom of the Muslim world. 'When the Muslim domination collapsed in Spain', remarks Figuier,[9] 'alchemy had already secured a new home on Western soil.' * The translations of this period were done into Latin, and this language was the usual medium in which the adepts embodied, or embedded, their observations and ideas during the next five centuries. Boyle's *Sceptical Chymist* (1661) marked alike the decline of alchemy and the gradual abandonment of Latin, in favour of their native tongues, by the exponents of the new chemistry.

The translators of the twelfth century were followed by the encyclopaedists of the thirteenth century, who played an important part in collecting, editing, and disseminating the knowledge which had come to them from the Muslim civilisation. One of the greatest of these scholastic philosophers was Albertus Magnus (1193–1282), who became known as *Doctor Universalis*, or 'the universal Doctor': his writings afford a valuable summary and exposition of the new knowledge, including alchemy. Great as was Albertus, the dominating position in thirteenth-century science is occupied by Roger

---

* 'Quand la domination arabe se trouva anéantie en Espagne, l'alchimie avait déjà conquis sur le sol de l'Occident une patrie nouvelle.'

Bacon (1214–1292), 'the wonderful Doctor' (*Doctor Mirabilis*), who was termed by Humboldt 'the greatest apparition of the Middle Ages', and by Sarton 'one of the greatest thinkers of all ages'. Roger Bacon was the first exponent of the inductive method in science, that is, of reasoning based upon accurate observation and experiment. His *Opus Majus* (1268) has been characterised as 'at once the Encyclopaedia and the Organon of the thirteenth century'. 'There are two methods of obtaining knowledge,' he wrote in the *Opus Majus*,* 'by argument and by experiment; argument makes conclusions and forces us to agree to them, but it does not make us feel certain or so remove suspicion that the mind rests in assurance of truth, unless this be also found by experience.'

The work of Albertus, Bacon, and contemporary writers such as Bartholomew the Englishman and Vincent de Beauvais is redolent of the lamp rather than of the laboratory: indeed, as stated above (p. 23), the contributions of the mediaeval philosophers and alchemists to the theory and practice of alchemy were insignificant. Nevertheless, the writings of Roger Bacon and his contemporaries exercised so wide and potent an appeal that in the fourteenth and fifteenth centuries numerous spurious works appeared under such names as ' Albertus Magnus ', ' Roger Bacon ', ' Raymond Lully ' (1225–1315), ' Arnold of Villanova ' (1240–1313), and others of accepted authority. Many of these writings dealt speciously with the ' wild joys ' of gold-making, and the invention of printing in the middle of the fifteenth century imparted a great impetus to their circulation. Thus, a mass of falsely ascribed literature confers exceptional difficulty upon the study of this imperfectly known period in the history of chemistry (*cf.* p. 56). It is not until the beginning of

---

* The original Latin text runs thus: ' Duo enim sunt modi cognoscendi, scilicet per argumentum et experimentum. Argumentum concludit et facit nos concedere conclusionem, sed non certificat neque removet dubitationem ut quiescat animus in intuitu veritatis, nisi eam inveniat via experientia '.[10]

the sixteenth century that firm ground is reached again in the writings of Paracelsus and the later iatro-chemists (p. 30); and even here the publications of the mysterious Basil Valentine (see Chapter V) tempt the unwary reader to follow still another false trail.

Hopkins [11] pictures the alchemical writers of the thirteenth century and later as ' imitators, receiving and copying the thoughts expressed by their predecessors and quoting at length what was to them and to mediaeval Europe unintelligible . . . advertised and exploited by charlatans, alchemy remained mediaeval while all other sciences advanced '.

§ 2. EXAMPLES OF ALCHEMICAL MANUSCRIPTS

Besides the almost incredible mass of printed alchemical literature which has come down to us, there are multitudes of surviving manuscripts, many of which have remained unprinted, and these are being systematically catalogued under the auspices of the Union Académique Internationale.[12] The British libraries are rich in alchemical manuscripts written in Latin and in English, but they contain relatively few in Greek. Latin alchemical manuscripts draw their material to some extent from Graeco-Egyptian sources; but, as Mrs D. W. Singer [12] has pointed out, ' Arabic alchemy was the main external factor in moulding Latin alchemy '. In general, the manuscript literature deals with practical workshop recipes as well as with mediaeval alchemy's main obsession of gold-making. The fine study of alchemical manuscripts is a task beset with difficulties; for the texts were continually altered as they passed from one alchemist to another, and the names of the alleged authors are often utterly misleading.

The *Speculum Secretorum Alchemiae* (' Mirror of the Secrets of Alchemy '), ascribed to Roger Bacon, may be quoted as a typical example of a simple Latin alchemical manuscript: six fourteenth-century and four fifteenth-

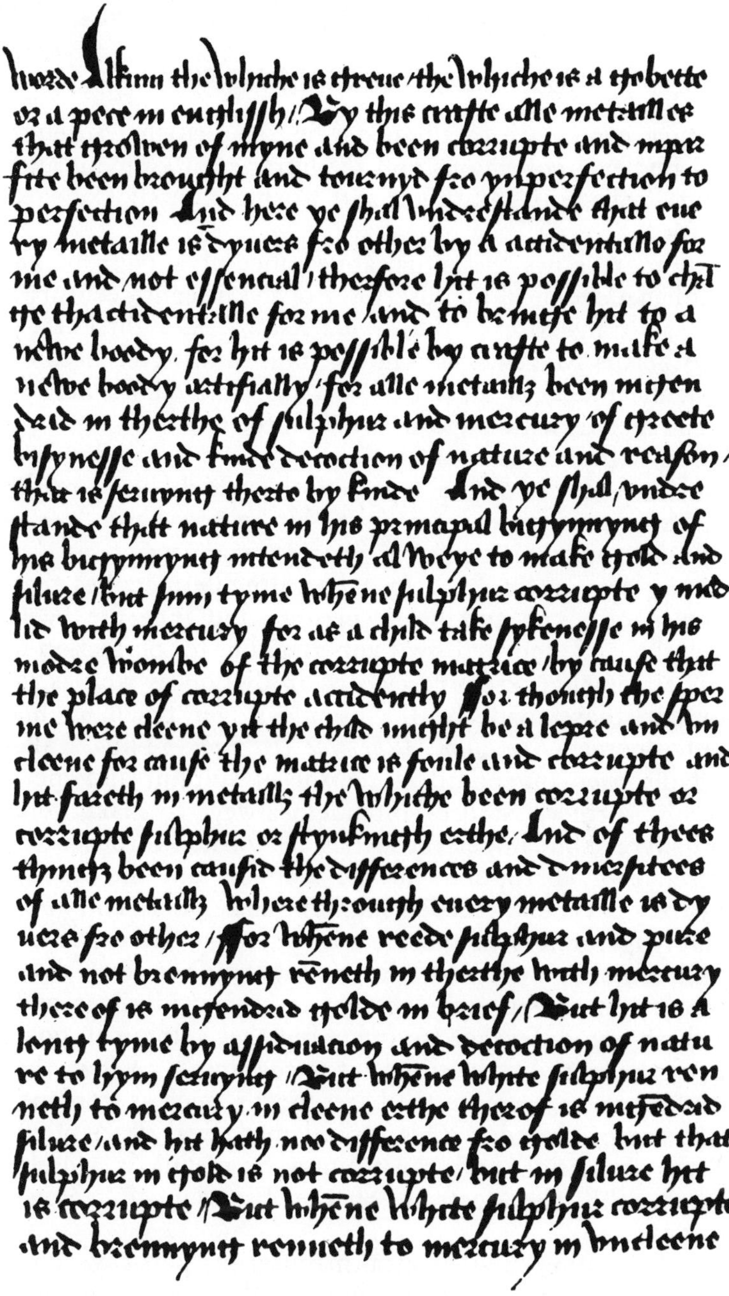

heede. Alkini the whiche is etrewe the whiche is a tobette
or a pece in englissh. By this crafte alle metalles
that throwen of myne and been corrupte and mar
fite been brought and torneyd fro ynperfection to
perfection. And here ye shall vndestande that eue
ry metalle is dyuers fro other by a accidentalle for
me and not essencial, therfore hit is possible to chā
ge the accidentalle forme and to be nese hit to a
newe boody, for hit is possible by crafte to make a
newe boody artifially, for alle metallz been nesen
deid in therthe of sulphur and mercury of greete
fynnesse and fmid deiocion of nature and reason,
that is seruyng therto by kinde. And ye shall vnde
stande that nature in his prinapil bigynnyng of
his bigynnyng intendeth alweye to make gold and
siluer, but sum tyme whene sulphur corrupte y med
lid with mercury for as a child take sykenesse in his
modre wombe of the corrupte mattier, by cause that
the place of corrupte accidently, for thought the sper
me were cleene yit the child might be a lepre and vn
cleene for cause the mattier is foule and corrupte and
hit fareth in metallz the whiche been corrupte or
corrupte sulphur or stynkmith erthe. And of these
thynges been caused the differences and diuersetees
of alle metallz, wherethrough euery metalle is dy
uere fro other, for whene reede sulphur and pure
and not brennyng reneth in therthe with mercury
there of is mesended golde in brief, but hit is a
lenty tyme by assiduacion and deiocion of natu
re to hym seruyng, but whene white sulphur ren
neth to mercury in cleene erthe thereof is mesedeid
siluer, and hit hath noo difference fro golde, but that
sulphur in golde is not corrupte, but in siluer hit
is corrupte, but whene white sulphur corrupte
and brennyng renneth to mercury in vncleene

FIG. 5.—A page of an English alchemical MS. of the 15th century.
Sloane MS. 353, f. 53v, British Museum (Albertus Magnus, *Semita Recta*).

45

century copies are catalogued [13] in the British Museum and Bodleian collections and the libraries of Gonville and Caius College, Cambridge, and Corpus Christi College, Oxford.

Fig. 5 shows the actual appearance of a page of an English alchemical manuscript [14] of the fifteenth century: this is a translation of the *Semita Recta* (' Straight Path ') of Albertus Magnus, containing interesting references to alchemical ideas of the period. Thus, ' mercury is modre of alle metalls with sulphur and with the rede stoone . . . and in nature he is hoote and moyste. . . . Sulphur is a fattenesse and a quyke myne of the which he is made thicke by temperate decoction til hit be very harde.'

' Alkymya ' is defined in this manuscript as ' a cunyng or a crafte . . . by this crafte alle metailles that growen of myne and been corrupte and inparfite been brought and tournyd fro ynperfection to perfection '. The statements resemble closely, in diction as well as in content, those made by the pseudo-Roger Bacon in his contemporary *Speculum Alchemiae* (p. 24). Thus, ' alle metaills been ingendard in therthe of sulphur and mercury '; further, Nature ' intendeth alweye to make gold and silvre ', but imperfect metals are formed through the presence of corruptions in the sulphur and the earthy matrix. ' Whene reede sulphur and pure and not brennyng renneth in therthe with mercury thereof is ingendrid golde . . . but whene white sulphur renneth to mercury in cleene erthe thereof is ingendrid silvre and hit hath noo difference fro golde but that sulphur in gold is not corrupte but in silvre hit is corrupte.' When the mercury also is impure, the baser metals, such as iron and lead, are produced. This typical mediaeval description of the formation and constitution of metals is closely linked with the doctrine of the Philosopher's Stone (p. 133).

The writing of alchemical manuscripts persisted long after the invention of printing, partly because of the importance of colour in the pictorial representations (*cf.*

p. 92). Sometimes, however, complete printed works were copied in manuscript, presumably because of the scarcity or cost of the printed original.* The marvellous craftsmanship displayed in much of this work bears witness to the skill and patience of the copyists and artists, and also to the amount of leisure which they enjoyed. The St. Andrews collection contains, for example, a complete manuscript copy, dating from the seventeenth century, of an edition of the *Novum Lumen Chymicum* (' New Light of Alchemy ') dated 1628, written in Latin in a fair hand; also, a manuscript addition to a printed copy of Khunrath's *Amphitheatrum* of 1609 reproduces in detail the many kinds and sizes of type used in the original.

A noteworthy Latin manuscript in the same collection, entitled *Philosophorum Praeclara Monita* (' The most renowned maxims of philosophers ') and written in French and Latin, is based on works ascribed to George Ripley, Arnold of Villanova, Raymond Lully, Nicolas Flamel, Eirenaeus Philalethes, Jean de Ré, and others. It was written in the first decade of the eighteenth century, and contains many boldly designed and brightly coloured drawings. One of the reproductions of a ' figure of Abraham the Jew' from *Philosophorum Praeclara Monita* is shown in Plate 5. An interesting feature of this manuscript is an alchemical account of the story of the Creation, with coloured illustrations of remarkable ingenuousness.[15] One of the most beautiful and artistic of all alchemical manuscripts is the copy of Salomon Trismosin's *Splendor Solis* in the Harley collection (p. 67).

§ 3. THE SUM OF PERFECTION

As an example of an early printed work on alchemy, we may select the *Summa Perfectionis* (' Sum of Perfec-

---

* Even so late as the eighteenth century, Fielding's Parson Adams, with his handsome income of twenty-three pounds a year, and a wife and six children, used a manuscript version of Æschylus, which he had transcribed with his own hand (*Joseph Andrews*, chap. xi).

tion '), a publication which has been called ' the main chemical text-book of mediaeval Christendom '.[16]   This, with certain associated Latin treatises of which no Arabic originals are known, has always appeared under the name of Geber.   In company with the pseudo-Roger Bacon's *Speculum Alchemiae* ('Mirror of Alchemy') and other works, including the *Tabula Smaragdina Hermetis* (' Emerald Table of Hermes '), these reputed writings of Geber were printed in such editions as those dated 1541 (Nuremberg) and 1545 (Berne).   The Geber texts here concerned can be traced back through the first printed edition of about 1481 to earlier manuscript versions. They were embellished in the Nuremberg and Berne editions with sixteen bold woodcuts of furnaces and other alchemical apparatus.   The title-page of the Berne edition is headed *Alchemiae Gebri Arabis philosophi solertissimi, Libri* (' The Books of the Alchemy of Geber, the most skilful Arabian philosopher ').   The items ascribed to Geber in the collection were translated into English in 1678 by Richard Russell, under the title: ' The Works of Geber, The Most Famous Arabian Prince and Philosopher '.

A critical examination of the works here ascribed to Geber has left their authorship in doubt.   Ruska [17] regards the *Summa Perfectionis* as a work written originally in Latin, in the early years of the fourteenth century, by an unknown author who attributed it to Geber in order to endow it with the authority of a renowned name.   This conclusion is based upon a comparison of the *Summa* with Geber's authenticated works; and the date is deduced from the fact that while the *Summa* is not mentioned in the authentic works of Albertus Magnus and Roger Bacon, quotations from it occur in the *Margarita Novella* of 1330 (p. 56).   The title of the *Summa* is curiously reminiscent of the fifth precept of Hermes (p. 54).

The opening of the *Summa* details the impediments lying in the path of the disciple of alchemy: these may be infirmities of the body or of the soul, such as ' decrepit old Age ' or a ' Brain repleat with many *Fumosities* ', to quote the words of Richard Russell. Besides being skilled in natural philosophy, ' the artist should be temperate and slow to anger, least he suddenly (through the force of rage) spoil and destroy his works begun. Likewise also, he must keep his money, least, when he is come near to the end of his magistery, his money being all spent, he be forced to leave the end uncompleated.' The writer adds wisely, ' this science agrees not well with a man poor and indigent, but is rather inimical and adverse to him '. He then deals in the manner of the schoolmen with ' the reasons of men denying this art, which are afterwards confuted '. Thus, to the objection ' Nature perfects metals in a thousand years; but how can you in your artifice of transmutation, live a thousand years, seeing that you are scarcely able to extend your life to an hundred? ' answer is made: ' What Nature cannot perfect in a very long space of time, that we compleat in a short space by our artifice: for art can in many things supply the defect of Nature '.

In discussing the composition of metals, the writer modifies the sulphur-mercury theory by introducing three natural principles of metals under the names sulphur, arsenick, and argent-vive. The allusion to sulphur as ' a fatness of the earth ' is suggestive of the later statement of Paracelsus that ' the life of metalls is a secret fatnesse, which they have received from sulphur '. Arsenick ' needs not be otherwise defined than sulphur ', save that it is a tincture of whiteness instead of redness. Argent-vive, or mercury, ' is a viscous water in the bowels of the earth. . . . It is also (as some say) the matter of metals with sulphur.' The writer gives much accurate detail of an observational and practical kind, the properties of the metals, for example, being described with much felicity: ' We signifie to the Sons of Learning, that tin is a metallick body, white, not pure, livid, and sounding little, partaking of little earthiness; possessing in its root harshness, softness, and swiftness of liquefaction, without ignition, and not abiding the cupel, or cement, but extensible under the hammer '.

The *Summa* contains excellent working descriptions of the chief practical operations of alchemy and of the nature and use of the various kinds of apparatus. *Sublimation* is accounted the most important operation: it is defined as ' the elevation of a dry thing by fire, with adherency to its vessel '. This and the following operations are clearly described with illustrative woodcuts.

In purification by *descension*, the material is heated on a perforated support, so that ' after its fusion it may descend through the hole thereof '. *Distillation* effects ' the purification of liquid matter from its turbulent feces ' through ' an elevation of aqueous vapours in their vessel '. Distillation is described from the water-bath as well as by direct heating. 'More subtile separation is made by distillation in water, than by distilling in ashes. This he knows to be true, who when he had distilled oyl by ashes, received his oyl scarcely altered into the recipient; but willing to separate the parts thereof, was by necessity forced to distil it by water. And then by reiterating that labour, he separated the oyl into its elemental parts; so that from a most red oyl, he extracted a most white and serene water, the whole redness thereof remaining in the bottom of the vessel. Therefore by this magistery . . . we may attain the oyl of every thing determinately, viz. of all vegetables, and of their like.'

' *Calcination* is the pulverisation of a thing by fire, through privation of the humidity consolidating the parts; *solution* is the reduction of a dry thing into water; *coagulation* is the reduction of a thing liquid, to a solid substance, by privation of humidity; *fixation* is the convenient disposing of a fugitive thing, to abide and sustain the fire; *ceration* is the mollification of an hard thing, not fusible unto liquefaction.'

That the writer of the *Summa* possessed a first-hand acquaintance with the operations he describes is manifest; he was also well acquainted with the properties of the known metals; and he closes the work, under the heading 'Of the Probations of Perfection', with a convincing display of metallurgical knowledge.

Some of the difficulties of the early publishers of printed alchemical works is brought out by a notice in Latin which appears at the end of the Berne edition (1545) of the *Summa* and other works. It runs as follows:

'Iohannes Petreius to the studious reader, greeting.

'You have now before you, studious reader, a few small works, but by no means commonplace, on the making of gold; if I learn that they have pleased you, I will, a little later, produce more, indeed the principal ones of this sort, of those which I have by me. But, since the copies are partly injured by age and mould, partly damaged by the ignorance of the copyists, to such an extent that they cannot be corrected and put right save by comparison of many copies, I request that if anyone has any such work, either in Latin or in German, he will courteously allow me to have it, so that by an extensive collation these writings may reach the hands of students in a more perfect form. Those who do so may count on no small thanks from aspirants after this art, and will receive several printed copies from me, along with their own copy undamaged, free of charge. These works are the ones I have on hand; the best of them I have marked with an asterisk.'

A list of more than thirty tracts follows, and the eleven starred items include works ascribed to Hermes, Mary the prophetess, Morienus, Alphidius, and Raymond Lully, also the *Margarita Novella* (p. 55) and *Turba Philosophorum* (p. 56).

§ 4. THE EMERALD TABLE OF HERMES

Alchemical writings of the mediaeval period teem with references and allusions to a series of thirteen precepts, which appear to have been viewed in general as a kind of supernatural revelation to the 'sons of Hermes' by the patron of their 'Divine Art'. The Precepts of Hermes were known also as the Emerald Table of Hermes, or the Smaragdine Table (*Tabula Smaragdina*). According to an oft-repeated legend, the original emerald slab, upon which the precepts were said to have been inscribed in Phoenician characters, was discovered in the tomb of Hermes by Alexander the Great; a variant of this product of the alchemical imagination states that a certain woman called Zara, sometimes identified with Sarah the wife of Abraham, took the Table from the hands of the dead Hermes in a cave near Hebron, some ages after the Flood. These legendary accounts are typical examples of the extravagant tone which was so often adopted in order to lend authority to alchemical assertions. They were, however, received with superstitious credence by the mediaeval alchemists, and as late as 1609 the mystical Khunrath (p. 81) included in his *Amphitheatrum* a large double-page illustration in which the Hermetic precepts are depicted as engraved upon an enormous rock, in the Latin and German tongues.

In later days, the nature of these accounts of the discovery of the Table gave rise to doubts concerning its reputed antiquity. According to Ferguson [18] the *Tabula Smaragdina* was first printed—together with the *Summa Perfectionis* and other tracts (p. 48)—in *De Alchemia*, at Nuremberg in 1541, after it had been known for at least three hundred years. Indeed, it is accompanied in the Nuremberg publication by a commentary written upon it by a certain Hortulanus, who has been identified by some with John Garland (1202–1252); according to Ruska, [19] however, the Hortulanus concerned here lived in the middle of the fourteenth century. In the Berne edition (1545) of the *Summa Perfectionis* and other works, the Latin version is printed under the heading: 'The Emerald Tablet of Hermes the Thrice-Great, concerning Chymistry. Translator unknown. The words of the secrets of Hermes, which were written on the Tablet of Emerald, found between his hands in a dark cave wherein his body was discovered buried.' Ferguson remarked in 1906 that everything concerning the Emerald Table ' remains a problem: its legendary and romantic discovery; the original language, for it is known in Latin only; its author, whether one of the several personages of the name of Hermes, or an anonymous writer who ascribes it to him to give it authority; its possible connection with so-called Hermetic writings of an earlier time; the subject with which it deals '. Ferguson added, ' the whole Hermes-legend forms a legitimate subject of inquiry ', and since his day it has received a good deal of attention from chemical historians.

That the Table was known in the thirteenth century is evident from an allusion to it in a genuine work of Albertus Magnus entitled *De Rebus Metallicis et Mineralibus* (' Concerning things metallic and mineral '). Among historians of chemistry who claimed a still greater antiquity for the Table, von Lippmann pointed out in 1919 that internal evidence suggested that the Latin text was a translation from the Greek, although a Greek original was unknown. This was the position when Holmyard,[20] in 1923, discovered a corrupt Arabic version of the text in a work ascribed to Jabir, bearing the title *Kitāb Usṭuqus al-'Uss al-Thānī* (' The Second Book of the Foundation '). The Table thus dates from at least four hundred years before Albertus. Holmyard adds—although Ruska [21] adopts a more cautious attitude—' and probably twelve hundred; its existence in a Greek form is rendered in the highest degree probable, and it must be acknowledged that in the *Tabula* we have one of the oldest alchemical fragments known '.

Although the precepts of Hermes Trismegistos have sometimes been dismissed as meaningless, they appear to offer one of the oldest statements of fundamental alchemical doctrine. The Emerald Table, whatever its source may have been, exerted a profound influence upon alchemical writings of the thirteenth century and later: for that reason it is reproduced here.[22]

PRELUDE TO CHEMISTRY

The Precepts of Hermes, engraved upon the Emerald Table

1. I speak not fictitious things, but that which is certain and true.
2. What is below is like that which is above, and what is above is like that which is below, to accomplish the miracles of one thing.
3. And as all things were produced by the one word of one Being, so all things were produced from this one thing by adaptation.
4. Its father is the sun, its mother the moon; the wind carries it in its belly, its nurse is the earth.
5. It is the father of perfection throughout the world.
6. The power is vigorous if it be changed into earth.
7. Separate the earth from the fire, the subtle from the gross, acting prudently and with judgment.
8. Ascend with the greatest sagacity from the earth to heaven, and then again descend to the earth, and unite together the powers of things superior and things inferior. Thus you will obtain the glory of the whole world, and obscurity will fly far away from you.
9. This has more fortitude than fortitude itself; because it conquers every subtle thing and can penetrate every solid.
10. Thus was the world formed.
11. Hence proceed wonders, which are here established.
12. Therefore I am called Hermes Trismegistos, having three parts of the philosophy of the whole world.
13. That which I had to say concerning the operation of the sun is completed.

These oracular pronouncements were held in superstitious veneration by the mediaeval alchemists, who appear indeed to have regarded them as the alchemical creed, or profession of faith in the Divine Art. The words were often endowed with a talismanic significance; they were to be found engraved upon the laboratory walls and interspersed throughout the writings of these assiduous searchers after magic stones and rejuvenating elixirs. The sentences were held to embody either the fundamental principles of alchemy or the secret of transmutation, according to the predilections of the commentator. The Table appears to carry veiled directions for preparing the Philosopher's Stone; but, as Ferguson drily remarks,

' its significance does not lie on the surface.   The man that runs cannot read it, nor, for that matter, the man who sits.'

The second and third precepts of Hermes are essentially a statement of the alchemical doctrine of the unity of all things (p. 12): in them we may discern a dark reflection of the view that all forms of matter have a common origin, and a common soul, or essence, which alone is permanent; the outward form, or body, is merely the temporary abode of the imperishable soul; substances are produced by evolutionary processes, and are capable of undergoing transmutation.   In the fourth precept, the Sun and Moon—gold and silver, or sulphur and mercury, are pictured as the sources of the Stone, which is also linked with the Aristotelian elements.   The eighth precept is suggestive of the kerotakis, or the later Vase of Hermes (p. 148), in which the Stone—' the father of perfection throughout the world '—was held to be prepared.

§ 5. THE NEW PEARL OF GREAT PRICE

In the year 1546 there was published at Venice an elegant duodecimo volume in Latin,[23] under the title *Pretiosa Margarita Novella*, or ' The New Pearl of Great Price '.   This is one of the early printed works on alchemy, and it was issued from the Aldine press with the sanction of Pope Paul III and the Venetian Senate.   It is now very difficult to find a copy of this work, partly on account of the beauty of its typography, which has attracted the notice of connoisseurs; but, beyond this, it was greatly prized by the adepts as a helpful compendium of alchemical knowledge.   Renouard remarks that this rare work almost always occurs in a dilapidated condition, since it was so subject to accidents near the furnaces of the adepts, among whom it was a great favourite.   The St. Andrews collection contains a copy of the Venetian edition[24] of 1557.   Possibly because it was concerned so closely with the Elixir of Life, this volume—having evaded the furnaces of the adepts—has kept its youth in a surprising way; its delicate, diminutive type is as fresh as that of a modern book, and the paper shows scarcely a stain, save at the edges. Yet this volume was twenty years old when van Helmont was born!

The *New Pearl* is a version of an introduction to alchemy written by Petrus Bonus of Pola in 1330, and edited more than two hundred years later for the Aldine press by Janus Lacinius of Calabria. It consists mainly of quotations from earlier works, and thus affords a valuable guide to the influence of the older alchemists on fourteenth-century alchemy. This is an important consideration, since, according to Ruska,[25] the fourteenth and fifteenth centuries offer a wider field for historical research than any other period in the history of chemistry. First among about twenty authors cited in the *New Pearl* stands the pseudo-Rhazes, with seventy references; the second place is shared by the *Sum of Perfection* (p. 47) and the *Turba Philosophorum*,* each with thirty references. In devoting equal attention to two works so radically different in character, Petrus Bonus appears to have held the balance between the undoubted value of the *Sum* and the venerable authority of the *Turba*.

Altogether, the *New Pearl* is a completely uncritical justification of alchemy based upon philosophical and metaphysical arguments. It expounds the history and principles of alchemy, the arguments for and against it, the nature of metals, the nature and operation of the Philosopher's Stone, and many other alchemical ideas, as culled from the reputed writings of Arnold of Villanova, Raymond Lully, Rhazes, Albertus Magnus, Michael Scot, and other authors whose names do not accompany the foregoing on the title-page. It contains some extraordinary statements, which must have caused even the

---

* The *Turba Philosophorum* is a collection of about 70 anonymous tracts, dating probably from the twelfth century, and held in great veneration by the later alchemists. The title may be rendered either as ' Conflict of the Philosophers ' or ' Convention of the Philosophers '. Pythagoras must be regarded as president of the ' convention ', but most of the imaginary speakers bear strange names.[26]

PLATE II

The Alchemical Microcosm and the Macrocosm.

From *Musaeum Hermeticum*, 1678.

PLATE 12

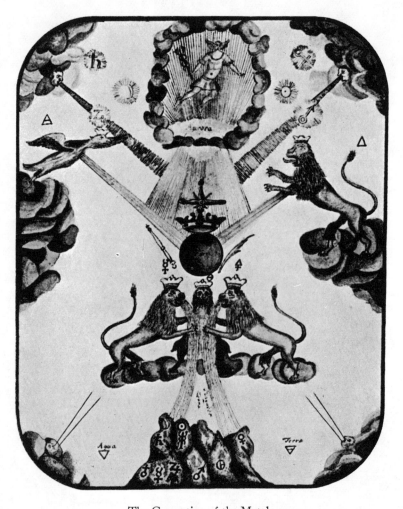

The Generation of the Metals.

From an 18th-century German MS. in the St. Andrews collection. (See p. 84.)

PLATE 13

Metals, Planets, and Signs of the Zodiac.

From de Bry's title-page of Maier's *Viatorium*, Oppenheim, 1618. A portrait
of Maier appears at the top. The associations are: gold, Sol, lion; silver,
Luna, crab; iron, Mars, ram, and scorpion; copper, Venus, bull, and scales;
tin, Jupiter, archer, and fishes; lead, Saturn, goat, and water-carrier; quicksilver,
Mercury, virgin, and twins.

PLATE 14

'Sow your gold in white earth.'

From *Secretioris Naturae Secretorum Scrutinium Chymicum*, Maier, 1687 (reprinted from *Atalanta Fugiens*, Maier, 1618). (See p. 95.)

Typicarum imaginum expofitio .

TRIA *in opere feruanda funt, primo materiá prǽpa=
ra, fecundo opus continua , ne difcontinuatione diffipetur ,
tertio fis patiens intima naturǽ paffim ueftigia feruans .*
▷ PRAEPARA *primo aquam uitǽ fummǽ purifica=
tam , & eá ferua . Non tamen credas , ut liquor ifte , quo
cunĉta madent , fit Bacchi candens ac limpidus humor . Nam
te dum uarijs immenfa per auia rebus , detines intentum,
felices prǽteris undas.*

PALATIVM *ingrediés in quo quindecim funt man
fiones ubi Rex diademate coronatus in excelfa fede fceptrum
totius orbis in manu tenens erit: coram cuius maieftate filius
cū quinq; famulis uarijs indutis ueftibus , flexis genibus regé
deprecantur ut tam filio quám feruis regnum impertiri di=
gnaretur , quorum precibus nil Rex ipfe refpondit .*

At films

FIG. 6.—A page of *Pretiosa Margarita Novella*, 1557.
The king (gold), his son (mercury), and five
servants (silver, copper, iron, tin, and lead).

57

most fire-hardened adept to reach for his salt-box. For instance: [27] ' Something closely analogous to the generation of Alchemy is observed in the animal, vegetable, mineral, and elementary world. Nature generates frogs in the clouds. . . . Avicenna tells us that a calf was generated in the clouds, amid thunder, and reached the earth in a stupefied condition. The decomposition of a basilisk generates scorpions. In the dead body of a calf are generated bees, wasps in the carcase of an ass, beetles in the flesh of a horse, and locusts in that of a mule. These generations depend on the fortuitous combination of the same elements by which the animal or insect is ordinarily produced.' Such statements throw a startling light upon alchemical modes of thought and reasoning.

This book contains some remarkable allegorical woodcuts. One of them offers an emblematic illustration of the four elements of Aristotle, in which earth, water, air, and fire are represented by a bear, a dragon, a bird, and an angel, respectively. Fourteen other woodcuts delineate an allegorical exposition of the various stages in the process of transmutation. A crowned king (gold) is approached by his son (mercury) and his five servants (silver, copper, iron, tin, and lead), who beseech him to change them into kings also (Fig. 6). The king maintains a diplomatic silence, whereupon he is killed by mercury. After passing through a series of remarkable vicissitudes, representing alchemical operations, the king rises from the dead and is at last able to accede to the original petition. The final woodcut shows a royal flush of crowned kings, from which, however, there is an unexplained absentee—possibly lead, which was always regarded with a certain amount of suspicion in alchemical circles. The narrative ends: ' Let no impostor, greedy or wicked person, touch this glorious work with his unclean hands. Let the honest man and him of a wise heart come hither, and him who is capable of exploring the most hidden causes of things.'

THE LITERATURE OF ALCHEMY

§ 6. THE FIGURES OF ABRAHAM THE JEW

Prominent in alchemical literature are certain writings ascribed to Nicolas Flamel (1330–1418). According to tradition, this Paris scrivener was the most celebrated of the French adepts and one of the most successful of the gold-makers. Evidences of Flamel's house, dating from 1407, are still to be seen in the fabric of No. 51 in the rue de Montmorency; and an inscribed marble tablet from his tomb in the old church of St. Jacques-la-Boucherie, now demolished, is preserved in the Musée de Cluny. This measures about 58 by 45 centimetres, with a thickness of 4 centimetres.[28] It is ornamented at the top with a carved representation of Christ, St. Peter, and St. Paul, containing interspersed symbols of the sun and moon. The tablet is shown in Plate 6. The main inscription * records that Nicolas Flamel, formerly a scrivener, left certain monies and properties for religious and charitable purposes, including gifts to churches and hospitals in Paris.

The tablet is of peculiar interest as constituting the sole original remnant of Flamel's famous ' hieroglyphics '. This unique relic of the heyday of alchemy has had a chequered career, for it was lost for many years after the demolition of the church of St. Jacques-la-Boucherie in 1797. It was finally run to earth in the shop of a greengrocer and herbalist in the rue des Arcis, who found that its smooth marble back formed an admirable chopping block for his cooked herbs!

Flamel exerted a profound influence upon mediaeval

---

* The wording is as follows: Feu Nicolas Flamel jadis escrivain a laissie par son testament a leuvre de ceste eglise · certaines rentes · et maisons · quil avoit acquestees · et achatees a son vivant · Pour faire · certain service divin · et distribucions dargent chascun an · par aumosne · touchans les quinze vins · lostel dieu et autres eglises et hospitaux de paris · Soit prie po les trespasses.

The scroll above the skeleton bears the words:
 Domine d[e]us in tua misericordia speravi.

The inscription at the bottom is as follows:
 De terre suis venus et en terre retourne
 Lame rens a toy Ihu qui les pechiez pardonne.

(*Ihu* is probably a contraction of *Jesus Hominum Ultor*.)

alchemy; according to Lacroix,[29] indeed, it was through him that the search for the Philosopher's Stone became the mania of the fifteenth century. This influence was due partly to his great reputation as a successful gold-maker, and partly to his picturesque use of symbolical figures: these were heraldic in their use of colour and capable of a dual interpretation in terms of theology and alchemy. The arcade in the churchyard of the Innocents, containing the original figures, formed a sacred shrine to the alchemists of succeeding centuries. Built in 1407, it survived until the eighteenth century. Much of the symbolism of mediaeval alchemy is derived from these celebrated figures, or 'hieroglyphics', and from Flamel's no less picturesque writings.

The following extracts are taken from a late English translation of a famous work [30] reputed to have been written by Flamel himself. It is entitled: 'Nicholas Flammel, His Exposition of the Hieroglyphicall Figures which he caused to bee painted upon an Arch in *St. Innocents Church-yard, in* Paris. *Faithfully, and (as the Maiesty of the thing* requireth) religiously done into English out *of the French and Latine Copies.* By Eirenævs Orandvs, *qui est, Vera veris enodans.* Imprinted at *London* by T. S. for *Thomas Walkley*, and are to bee solde at his Shop, at the Eagle and Childe in *Britans Bursse.* 1624.' The translator, apparently a 'veiled master' of the school of Basil Valentine (p. 183), seeks to enhance the effect of his anonymity by giving an anagram of his pseudonym, signifying 'unravelling truths by means of truths'.

'Although that I *Nicholas Flammel*, Notary, and abiding in *Paris*, in this yeere one thousand three hundred fourescore and nineteene, and dwelling in my house in the street of Notaries, neere unto the Chappell of St. *James* of the *Bouchery*; although, I say, that I learned but a little Latine, because of the small meanes of my Parents, which neverthelesse were by them that envie me the most, accounted honest people; yet by the grace of God, and the intercession of the blessed Saints in *Paradise* of both sexes, and principally of Saint *James* of *Gallicia*, I have not wanted the understanding of the Bookes of the *Philosophers*, and in them learned their so hidden secrets. . . .

' Whilest therefore, I *Nicholas Flammel, Notary*, after the decease of my Parents, got my living in our Art of Writing, by making *Inventories*, dressing accounts, and summing up the Expences of *Tutors* and *Pupils*, there fell into my hands, for the sum of two Florens, a guilded Booke, very old and large; It was not of Paper, nor Parchment, as other Bookes bee, but was onely made of delicate Rindes (as it seemed unto me) of tender yong trees: The cover of it was of brasse, well bound, all engraven with letters, or strange figures. . . .

' It contained thrice seven leaves, for so were they counted in the top of the leaves, and always every seventh leafe was without any writing, but in stead thereof, upon the first seventh leafe, there was painted a *Virgin*, and *Serpents* swallowing her up; In the second seventh, a *Crosse* where a *Serpent* was crucified [Plate 7, VI]; and in the last seventh, there were painted *Desarts*, or *Wildernesses*, in the middest whereof ran many faire fountaines, from whence there issued out a number of *Serpents*, which ran up and downe here and there [Plate 7, VII]. Upon the first of the leaves, was written in great Capitall Letters of gold, ABRAHAM THE JEW, PRINCE, PRIEST, LEVITE, ASTROLOGER, AND PHILOSOPHER, TO THE NATION OF THE JEWES, BY THE WRATH OF GOD DISPERSED AMONG THE GAULES, SENDETH HEALTH. After this it was filled with great execrations and curses (with this word MARANATHA, which was often repeated there) against every person that should cast his eyes upon it, if hee were not *Sacrificer* or *Scribe*. . . .

' The fourth and fifth leafe therefore, was without any writing, all full of faire figures *enlightened*, or as it were *enlightened*, for the worke was very exquisite. First [Plate 7, I] he painted a *yong man*, with wings at his anckles, having in his hand a *Caducæan* rodde, writhen about with two *Serpents*, wherewith hee strooke upon a helmet which covered his head; he seemed to my small judgement, to be the God *Mercury* of the *Pagans*: against him there came running and flying with open wings, a great old man, who upon his head had an *houre-glasse* fastened, and in his hands a hooke (or sithe) like *Death*, with the which, in terrible and furious manner, hee would have cut off the feet of *Mercury*.

' On the other side of the fourth leafe [Plate 7, II, also Plate 5], hee painted a faire *flowre* on the top of a very high *mountaine*, which was sore shaken with the *North wind*; it had the foot *blew*, the flowres *white* and *red*, the leaves shining like fine *gold*: And round about it the *Dragons* and *Griffons* of the *North* made their nests and abode.

' On the fifth leafe [Plate 7, III] there was a faire *Rose-tree* flowred in the middest of a sweet *Garden*, climbing up against a hollow *Oake*; at the foot whereof boyled a fountaine of most *white water*, which ranne head-long downe into the depths, notwithstanding it first passed among the hands of infinite people, which digged in the Earth seeking for it; but because they were blinde, none of them knew it, except here and there one which considered the *weight*.

' On the last side of the fift leafe [Plate 7, IV], there was a *King* with a great *Fauchion*, who made to be killed in his presence by some *Souldiers* a great multitude of little *Infants*, whose Mothers wept at the feet of the unpittifull *Souldiers*: the bloud of which *Infants* was afterwards by other Souldiers gathered up, and put in a great vessell, wherein the *Sunne* and the *Moone* came to bathe themselves. And because that this History did represent the more part of that of the *Innocents* slaine by *Herod* . . . I placed in their *Churchyard* these *Hieroglyphick Symbols* of this secret science.'

Flamel refers here to the elaborate frescoes, sometimes known as ' the figures of Abraham the Jew ', with which he decorated the arcade (p. 60) in the churchyard of the Innocents. Although, most unfortunately, these historic frescoes have perished, a detailed representation of the design has come down to us (Plate 7).[31]

The emblems occurring in the figures of Abraham pervade the later alchemical literature, and they were repeated age after age with surprisingly little variation. Thus, the fair flower growing on the windy hill, haunted by dragons and griffins, is shown in a spirited coloured drawing of the seventeenth century in the keeping of the Bibliothèque Nationale;[32] a still later version of the same figure, from the manuscript entitled *Philosophorum Praeclara Monita* (p. 47), is reproduced in Plate 5.

The white and red flowers of this picture represent the white and red stages of the Great Work, the red one being identical with Ben Jonson's ' flower of the sun, the perfect ruby, which he calls elixir '. The dragon is sophic mercury; the griffin, a combination of lion and eagle, that is, of the fixed and the volatile. The old man with the scythe, representing Saturn or Kronos (p. 91), cutting off the feet of Mercury, signifies the fixing of mercury. The adepts identified sophic mercury with the ' essence ' of silver (p. 25). Now when silver is cupelled with lead, its original impurities sink into the material of the cupel, and the residual silver becomes ' fixed ', or unalterable. Thus the pure ' essence ' of silver, or of quicksilver (regarded as a baser form of silver), has been obtained: Saturn has cut off the feet of ordinary mercury, or quicksilver, and rendered it immobile. The fountain of heavy water is the Hermetic stream, symbolic of sophic mercury (p. 99). As Flamel eventually discovered, to his great relief, infants' blood signified merely the mineral spirit of metals, ' principally in the *Sunne, Moone,* and *Mercury* ', that is, in gold, silver, and mercury. He relates that the strange language of the ' guilded Booke ' ' was the cause, that during the space of *one and twenty yeeres,* I tryed a thousand broulleryes, yet never with *bloud,* for that was wicked and villanous '.

Flamel describes [33] the colours of the figures with great care, as will be seen from the following accounts of

the central panels: ' The figure of a man, like that of Saint
Paul, cloathed with a robe white and yellow, bordered
with gold, holding a naked Sword, having at his feet a man
[Flamel] on his knees, clad in a robe of orange colour,
blacke and white, holding a roule. . . . Upon a green
field, three resusitants, or which rise againe, two men and
one woman, altogether white: Two Angels beneath, and
over the Angels the figure of our Saviour comming to
judge the world, cloathed with a robe which is perfectly
Citrine white. . . . The figure of a man, like unto Saint
Peter, cloathed in a robe Citrine red, holding a key in
his right hand, and laying his left hand upon a woman
[Perrenelle], in an orange coloured robe, which is on her
knees at his feete, holding a Rowle.'

In another place [34] Flamel shows that he attached
importance to the sequence (*cf.* p. 16) of colours in the
operations of the Great Work, when he writes as follows:
' I have also set against the wall, on the one and the other
side, a *Procession*, in which are represented by order all
the colours of the *stone*, so as they come and goe, with this
writing in French:

> *Moult plaist a Dieu procession,*
> *S'elle est faicte en devotion:*

that is,

> *Much pleaseth God procession,*
> *If't be done in devotion.*'

Flamel's discovery of the mysterious ' guilded Booke '
led him to spend much time in vainly seeking its meaning,
which, as he says, ' made me very heavy and sollitary, and
caused me to fetch many a sigh. My wife *Perrenelle*,
whom I loved as my selfe, and had lately married, was
much astonished at this. . . . I could not possibly hold
my tongue, but told her all . . . whereof . . . she
became as much enamored as my selfe.' At last, after
fruitlessly consulting ' the greatest Clerkes in *Paris* ', he
provided himself with copies of the pictures, and ' with
the consent of *Perrenelle* . . . having taken the *Pilgrims*
habit and staffe ', he journeyed into Spain, in the hope of

finding in that ancient home of alchemy an informed Jewish priest. Flamel's adventures in search of the secret of the Stone constitute one of the epics of alchemy. The climax came at Leon. Here, a Jewish physician called Master Canches was ' ravished with great astonishment and joy' at the sight of the ' enigmaes ', which he began to decipher forthwith. This Master Canches, ' who was very skilfull in sublime Sciences ', was indeed so ' transported with great Ardor and joy ' that he agreed to accompany Flamel back to Paris. Part of the journey was by sea, across the tip of the redoubtable Bay of Biscay; and to Flamel's great grief, his companion, ' being afflicted with excessive vomitings, which remained still with him of those he had suffered at Sea ', died at Orleans.

Flamel journeyed sadly on alone; but sorrow took wing when he reached Paris and came within sight of the humble little house, bearing the sign of the Fleur-de-Lys —at the corner of the rue des Écrivains and the rue de Marivaus. ' He that would see the manner of my arrivall, and the joy of *Perenelle*, let him look upon us two, in this *City* of *Paris*, upon the doore of the Chappell of *St James* of the *Bouchery*, close by the side of my *house*, where wee are both painted, my selfe giving thankes at the feet of *Saint James of Gallicia*,* and *Perrenelle* at the feet of *St John*, whom shee had so often called upon.' Flamel here refers to a small decorated portal which he erected in 1389 at the adjacent church of St. Jacques-la-Boucherie.

According to his account, Flamel was now armed— thanks to the unfortunate Master Canches, of melancholy memory—with a knowledge of ' the first *Principles*, yet not their first *preparation*, which is a thing most difficult, above all the things in the world: But in the end I had that also, after long errours of *three yeeres*, or thereabouts; during which time, I did nothing but study and labour. . . . Finally, I found that which I desired, which I also soone knew by the strong *sent* and *odour* thereof. Having this, I easily accomplished the *Mastery*, for knowing the

* Whose shrine Flamel had visited during his pilgrimage to Spain.

preparation of the first *Agents*, and after following my Booke according to the *letter*, I could not have missed it, though I would.

' Then the first time that I made *projection*, was upon *Mercurie*, whereof I turned halfe a pound, or thereabouts, into pure *Silver*, better than that of the *Mine*, as I my selfe assayed, and made others assay many times. This was upon a Munday, the 17. of *January* about noone, in my house, *Perrenelle* onely being present; in the yeere of the restoring of mankind, 1382. And afterwards, following alwayes my Booke, from word to word, I made *projection* of the *Red stone* upon the like quantity of *Mercurie*, in the presence likewise of *Perrenelle* onely, in the same house, the *five and twentieth day* of *Aprill* following, the same yeere, about five a *clocke* in the *Evening*; which I transmuted truely into almost as much pure *Gold*, better assuredly than common *Golde*, more soft, and more plyable. I may speake it with truth, I have made it three times, with the helpe of *Perrenelle*, who understood it as well as I, because she helped mee in my operations, and without doubt, if shee would have enterprised to have done it alone, shee had attained to the end and perfection thereof. I had indeed enough when I had once done it, but I found exceeding great pleasure and delight, in seeing and contemplating the *Admirable workes of Nature*, within the *Vessels*.'

This early circumstantial account of an alleged transmutation by projection has a graphic and verisimilous style which inevitably brings to mind certain later narrations of successful gold-making, in particular that of Helvetius,[35] describing his alleged transmutation of lead into gold at the Hague, in 1666. In ending his story, Flamel refers to his numerous benefactions, and exhorts others who may ' conquere this *rich golden* Fleece ' to go and do likewise. ' At that time when I wrote this *Commentarie*,' he says, ' in the yeere *one thousand foure hundred and thirteene*, in the end of the yeere, after the decease of my faithfull companion, which I shall lament all the dayes of my life: she and I had already founded, and endued with revenewes 14. *Hospitals* in this *Citie* of *Paris*, wee had new built from the ground *three Chappels*, we had inriched with great gifts and good rents, *seven Churches*, with many reparations in their *Churchyards*, besides that which we have done at *Boloigne*, which is not much lesse than that which wee have done heere.'

The 'guilded Booke', the allegorical paintings, the alleged transmutations, and Flamel's rise to affluence have provided matter for many discussions and hypotheses.[36] His wealth has been ascribed to professional activities, fortunate speculations, inheritances from proscribed Jews, and other conventional sources; but whatever the facts may have been, the story of Nicolas Flamel and his devoted Perrenelle will go down through the ages as one of the most romantic episodes in the history of alchemy.

§ 7. THE SPLENDOUR OF THE SUN

There is in the British Museum a vellum manuscript version of a tract entitled *Splendor Solis* ('Splendour of the Sun'), the authorship of which has been ascribed to one Salomon Trismosin. This Harley MS. 3469, of date 1582, is a striking example of an alchemical manuscript of the late mediaeval period. It consists of a German text of forty-eight leaves, illuminated in gold, with twenty-two coloured allegorical paintings of great beauty, representing alchemical ideas and processes. The second of these paintings depicts an adept carrying a liquid in a Hermetic Vase, from the neck of which flows a ribbon with the inscription 'Eamus Quesitum Quatuor Elementorum Naturas' ('Let us go to seek the natures of the four elements'). In the background is a fair landscape. The design is enclosed within a golden border, decorated with a butterfly, a bee, flowers, red currants, an owl, a peacock, and other birds, together with a stag and a hind. Among the familiar emblems depicted in the paintings are the king and queen, the drowning king, the death and resurrection of the king, and a hermaphrodite holding a golden egg. The Philosophic Tree is shown with birds and an issuing Hermetic spring; its fruit is being gathered by a man who has ascended a ladder with seven rungs. Seven of these pictures illustrate processes in the preparation of the Stone, and are described in Chapter III (p. 150): one of them is reproduced as the frontispiece of this book.

This tract was published at Rorschach in 1598, at Basel in 1604, and again at Hamburg in 1708, in a printed collection of tracts entitled *Aureum Vellus oder Guldin Schatz und Kunstkammer* (' The Golden Fleece, or Golden Treasure and Art-Chamber '). The woodcuts in these editions are simplified and crude reproductions of the illustrations of the Harley manuscript; some of them are plain, others are hand coloured. The tract, written in allegorical and somewhat extravagant language, is a characteristic production of the esoteric type. ' Unaided Nature ', states the writer,[37] ' does not produce things whereby imperfect metals can in a moment be made perfect, but by the secrets of Our Art this can be done . . . although the . . . Stone can only be brought to its proper form by Art, yet the form is from Nature.'

The Great Work of preparing the Stone is ' explained by a few suitable illustrations, parables, and various aphorisms of the philosophers '. The veiled comments on the pictures are full of allusions to the reputed sayings of Hermes, Avicenna, Aristotle, Menaldus, Senior (p. 156), Rosinus, Ovid, Socrates, Rhazes, Alphidius, Morienus, and other sages, real and imaginary. Thus, the following cryptic remark of Senior is quoted: ' Our Fire is a Water. If you can give a Fire to a Fire and Mercury to Mercury, then you know enough.'

Much emphasis is laid upon the colours which are said to appear in the preparation of the Stone. It is pointed out, for example, that ' Miraldus, the Philosopher, says in the *Turba*: It turns black twice, yellow twice, and red twice . . . the principal colours are black, white and red; between these many others appear'. It is apparently for this reason that so much importance is attached to the colouring of the illustrations in alchemical manuscripts. In this respect, the alchemical manuscript retained an advantage over the printed book; sometimes paintings or hand-coloured drawings were pasted into the printed versions. *Splendor Solis* ends by describing alchemy as 'the most noble Art and comforter of the poor, above all natural arts, which man may ever have on earth'.

Kopp, one of the celebrated historians of chemistry, questioned the existence of such a person as Salomon Trismosin, and concluded that the tracts published under his name were spurious. It was possibly for the purpose of anticipating such criticisms that in the forefront of the *Aureum Vellus* was placed an account of Trismosin's wanderings and adventures in search of the Philosopher's Stone: this is written with a lightness of touch and a wealth of circumstantial detail which certainly impart an air of verisimilitude to the bald and unconvincing narrative of the *Splendor Solis*. It is of interest as a typical description of the operations of pseudo-alchemists and the eventual discovery of the Philosopher's Stone by the hero.[37]

'When I was a young fellow,' it begins, 'I came to a miner named Flocker, who was also an Alchemist, but he kept his knowledge secret, and I could get nothing out of him. He used a process with common lead, adding to it a peculiar sulphur, or brimstone, he fixed the lead until it became hard, then fluid, and later on soft like wax.

'Of this prepared lead, he took 20 loth (10 ounces), and 1 mark pure unalloyed silver, put both materials in flux and kept the composition in fusion for half an hour. Thereupon he parted the silver, cast it in an ingot, when half of it was gold.

'I was grieved at heart that I could not have this art, but he refused to tell his secret process.

'Shortly thereafter he tumbled down a mine and no one could tell what was the artifice he had used.

PRELUDE TO CHEMISTRY

' As I had seen it really done by this miner, I started in the year 1473 on my travels to search out an artist in Alchemy, and when I heard of one I went to him, and in these wanderings I passed 18 months, learning all kinds of alchemical operations, of no great importance, but I saw the reality of some of the *particular* processes, and I spent 200 florins of my own money, nevertheless I would not give up the search. I thought of boarding with some of my friends, and took a journey to Laibach, thence to Milan, and came to a monastery. There I heard some excellent lectures and served as an assistant, for about a year.

' Then I travelled about, up and down in Italy, and came to an Italian tradesman, and a Jew, who understood German. These two made English tin look like the best fine silver, and sold it largely. I offered to serve them. The Jew persuaded the trader to take me as a servant, and I had to attend the fire; when they operated with their art I was diligent, and they kept nothing from me, as I pleased them well. In this way I learnt their art, which worked with corrosive and poisonous materials, and I stopped with them fourteen weeks.

' Then I journeyed with the Jew to Venice. There he sold to a Turkish merchant forty pounds of this silver. While he was haggling with the merchant I took six loth of the silver, and brought it to a goldsmith, who spoke Latin, and kept two journeymen, and I asked him to test the silver. He directed me to an assayer on Saint Mark's Place, who was portly and wealthy. He had three German assay-assistants. They soon brought the silver to the test with strong acids, and refined it on the cupel; but it did not stand the test, and all flew away in the fire. And they spoke harshly to me asking where I got the silver. I told them I had come on purpose to have it tested, that I might know if it was real silver.

' When I saw the fraud, I returned not to the Jew, and paid no more attention to their art, for I feared to get into trouble together with the Jew, through the false silver.

' I then went to a College in Venice, and asked there if they could give me two meals daily while I looked for employment. The Rector told me of a Hospital where there were other Germans, and there we got sumptuous food. It was an institution for destitute strangers, and people of all nations came there.

' The next day I went to Saint Mark's Place, and one of the assay-assistants came up, and asked me where I got that silver ? Why I had it tested, and if I had any more of it? I said I had no more of that silver, and that I was glad to have got rid of it, but I had the art and I should not mind telling it to him. That pleased the assayer, and he asked me if I could work in a laboratory? I told him I was a Laborant travelling on purpose to work in alchemical laboratories. That pleased him vastly, and he told me of a nobleman who kept a laboratory, and who wanted a German assistant. I readily accepted, and he took me straight to the chief chemist [der Obrist Laborant], named Hans Tauler, a German, and he was glad to get me. So he engaged me on the spot at a weekly wage of two crowns and board as well [zwo Kronen und Essen darzu]. He took me about six Italian miles out of Venice to a fine large mansion called Ponteleone [Pontilon]. I never saw such laboratory work [Laborieren], in all kinds of Particular Processes, and medicines, as in that place. There everything one could think of was provided and ready for use. Each workman (of whom there were nine) had his own private room, and there was a special cook for the whole staff of Laborants.

' The chief chemist gave me at once an ore to work on, which had been sent to the nobleman, four days previously. It was a cinnabar [coagulatio Mercurii] the chief had mixed with certain powders, just to try my knowledge, and he told me to get it done in two days. I was kept busy, but succeeded with the Particular Process, and on testing the ingot of the fixed Mercury, the whole weighed nine loth, the test gave three loth of fine gold.

' That was my first work and stroke of luck. The

chief chemist reported it to the nobleman, who came out unexpectedly, spoke to me in Latin, called me his Fortunatum, tapped me on the shoulder and gave me twenty-nine crowns. He spoke a funny kind of Latin I could hardly understand, but I was pleased with the money.

'I was then put on oath not to reveal my Art to anyone. To make a long story short, everything had to be kept secret, as it should be. If someone boasts of his art, even if he has got the Truth, God's Justice will not let such a one go on. Therefore be silent [halt das Maul], even if you have the highest Tincture, but give charity.

'I saw all kinds of operations at this nobleman's laboratory, and as the chief chemist favoured me, he gave me all kinds of operations to do, and also mentioned that our employer spent about 30,000 crowns on these arts, paying cash for all manner of books in various languages, to which he gave great attention. I myself witnessed that he paid 6000 crowns for the manuscript *Sarlamethon*, a process for a Tincture in the Greek language. This the nobleman had soon translated and gave me to work. I brought that process to a finish in fifteen weeks. Therewith I tinged three metals, copper, tin and silver, into fine gold; and this was kept most secret.

'This nobleman was gorgeous and powerful, and when once a year the Signoria went out to sea, to witness the throwing of a gem ring into the water at the ceremony of wedding the Adriatic, our gentleman with many others of the Venetian nobility went out in his grand pleasure ship, when suddenly a hurricane arose and he, with many others of the Venetian lords and rulers, was drowned. The laboratory was then shut up by the family and the men paid off, but they kept the chief chemist.

'Then I went away from Venice, to a still better place for my purpose, where cabbalistic and magical books in the Egyptian language were entrusted to my care; these I had carefully translated into Greek, and then again re-translated into Latin. There I found and captured the Treasure of the Egyptians. I also saw what was the great

PLATE 15

Astrological Schemes from Norton's *Ordinall*.

From *Theatrum Chemicum Britannicum*, Ashmole, 1652.

PLATE 16

The Hermetic Androgyne.

From the MS. *Philosophorum Praeclara Monita* (p. 47). The sun-tree is shown on the left, the moon-tree on the right.

PLATE 17

The Philosopher's Egg.

From *Secretioris Naturae Secretorum Scrutinium Chymicum*, Maier, 1687 (reprinted from *Atalanta Fugiens*, Maier, 1618).

PLATE 18

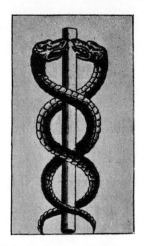

(i) A simple form of Caduceus.

From *Nicolai Flamelli, Chymische Werke*, Wienn, 1751.
An Egyptian form, almost identical with the
above, is of predynastic origin. (See p. 106.)

(ii) The Vitriol Acrostic.

From *Viridarium Chymicum*, Stolcius, 1624. (See p. 104.)

subject they worked with, and the ancient heathen kings used such Tinctures and have themselves operated with them, namely Kings *Xofar, Sunsfor, Xogar, Xopholat, Iulation, Xoman* and others. All these had the great treasures of the Tincture, and it is surprising that God should have revealed such secrets to the heathen, but they kept it very secret.

' After a while I saw the fundamental principles of this art, then I began working out the best Tincture (but they all proceed, in a most indescribable manner from the same root). When I came to the end of the Work I found such a beautiful red colour as no scarlet can compare with, and such a treasure as words cannot tell, and which can be infinitely augmented. One part tinged 1500 parts of silver into the best gold. . . . I will not tell how after manifold augmentation what quantities of silver and other metals I tinged after the multiplication. I was amazed. . . .

> ' Study what thou art,
> Whereof thou art a part,
> What thou knowest of this art,
> This is really what thou art.
> All that is without thee
> Also is within.   Amen.' *

This attractive narrative, whatever its source, throws much light upon the habits and practices of the itinerant ' labourer in the fire ', or ' laborant ', of mediaeval times. Despite the considerable traffic in false alloys simulating gold and silver, it is clear from the narrative that there were competent assayers, even in those days, who were not deceived by these ' colourable imitations '. It is unfortunate that the writer omitted to take his own ' best

* The original German runs as follows in the Hamburg edition of 1708 (p. 5):

> ' Studier nun darauss du bist/
> So wirst du sehen was da ist.
> Was du studierst/ lehrnest und ist/
> Das ist eben darauss du bist.
> Alles was ausser unser ist/
> Ist auch in uns/ Amen.'

gold ', prepared by transmutation from silver, to the
' portly and wealthy ' Venetian assayer whose assistants
detected so promptly the spurious nature of the silver
made from English tin! Another interesting point
brought out in this account is the expensiveness of manuscripts and books dealing with the secret processes of
gold-making. Demand creates supply; and the great
demand for such writings during the age of pseudo-
alchemy was probably one of the chief causes of the flood
of alluring and pseudepigraphic literature of the kind.

§ 8. The Practical Tradition in Alchemy

If alchemy be regarded as the child of Greek philosophy, its other and older parent is still to be sought in the
practical knowledge of Egypt and other ancient civilisations (p. 8). The practical tradition was maintained
during the Islamic era, and at the Renaissance it passed
in an attenuated form into Europe. The records of
the industrial arts of the ensuing epoch are regrettably
scanty, although, as Lacroix[38] remarks, 'more profit would
have been derived from consulting the daily note-book
of an artisan of that period than the farrago of those
who were engaged in the Great Work '. That much
sound knowledge of chemical materials and processes
existed in Europe during the so-called period of pseudo-
alchemy is evident from a study of some of the relevant
writings which remain.

The manufacture of gunpowder, for example, called
for the effective application of such knowledge. The discovery of this marvel of the Middle Ages is usually attributed to Roger Bacon: its composition is stated—partly
concealed in an anagram, of which the meaning and
authenticity have been contested—in his tract *De Mira-
bili Potestate Artis et Naturae* (' On the marvellous power
of art and nature '), written in 1242. Tradition ascribes
the discovery, in the following century, of its propellent
power to another monk, Berthold Schwarz, of Freiburg im

Breisgau.   Several old engravings show 'Black Berthold'
—whose real name was probably Constantin Anklitzen—
in the act of firing gunpowder, by means of a flint and
steel, in a mortar, from which the pestle (or, in some
engravings, a wooden block) is being projected into
the air.

The *Codex Germanicus*, dating from about 1350, gives
some interesting practical details of the methods in use
at that time for purifying and testing, among other
materials, the constituents of gunpowder.   Saltpetre was
purified by crystallisation from a filtered solution, followed
by fusion to expel the last traces of water.   A curious but
reliable criterion was used in testing the purity of sulphur:
this is illustrated in Plate 8.[39]   The explanation runs as
follows: ' If thou wilt try whether sulphur be good or
not, take a lump of sulphur in thine hand and lift it to
thine ears.   If the sulphur crackle, so that thou hearest it
crackle, then it is good; but if the sulphur keep silent
and crackle not, then it is not good, and must be treated
as thou shalt hear hereafter how it shall be prepared.' *
Pure sulphur does indeed crackle when submitted to this
treatment, but if it contain as little as one per cent of
impurities it remains silent.

Among the early printed books dealing with severely
practical aspects of alchemy are Brunschwick's *Buch zu
Distillieren* (' Book of Distillation ') and Agricola's *De Re
Metallica* (' On Metallurgy ').   The first of these was one
of the earliest books of applied chemistry to appear in the
German language.   It contains a comprehensive account
of the process of distillation and its applications.   The
numerous bold woodcuts with which it is adorned were
the first attempts at representing chemical apparatus
and manipulations in this way.   The *Grosses Distil-
lierbuch* appeared in 1512; the modified edition of

* ' Wiltu Sweuel versuchen ob er gut sey oder nicht So nym ein knolln
sweuel in die hant / und heb in zu den oren krachent denn der Swebel daz du
in hörst krachen so ist er gut sweigt der swebel aber still und nicht kracht / so
ist er nicht gut so muz man in machn als du hie hernoch wol horn wirst / wie
man in beraitn sol.'

1519 is a noble volume containing 330 folio sheets, beautifully printed in Gothic letter, in double columns. This rare work is sometimes encountered in the original

Fig. 7.—The Rosenhut.
From the *Buch zu Distillieren*, Brunschwick, 1519.

German pigskin binding, richly blind-tooled, with brass bosses and clasps. ' Laus deo et honor ', wrote the author as he laid down his quill, and the printer added (p. 328): ' Here endeth this book happily printed and brought to a close in the Imperial city of Strassburg by Johann Grüninger on the eve of St. Adolphus in the year of our Lord 1519 '.

Fig. 7 reproduces a passage from this edition. The woodcut shows a ' Rosenhut '—or primitive form of air condenser—in a corner of the laboratory affording charming glimpses of the mediaeval city without. The text outlines the preparation of ' another good water for the palsy, not costly '. The procedure is stated in simple and concise language, which forms a refreshing contrast to the verbose and meaningless outpourings of mystical alchemy: ' Take of parsley seed, three ounces; of green wormwood, two good handfuls; of distilled wine, three ounces. Pound together, and distil either in an alembic or an ordinary Rosenhut as shown here. And of the water drink two spoonfuls every morning, fasting; rub the limbs with it, morning, noon and night, and let it dry in.'

De Re Metallica, the standard mediaeval work on mining and metallurgy, was first published at Basel in 1556. The author, George Bauer (Agricola), was physician to the miners of Joachimstal. It is a handsome book, of folio size, profusely illustrated with woodcuts; these afford an admirable idea of the processes and machinery described in the text, and at the same time they are spirited and artistic. The contents are not wholly metallurgical, as there are sections on assaying, cupellation, and the preparation of salt, nitre, and other common substances. Plate 9 is one of a series of woodcuts illustrating the manufacture of common salt: the original engraving measures about $9\frac{1}{4}$ by $5\frac{1}{2}$ inches. The practical and precise nature of the description may be gauged from a short extract: [40] ' Salt-water is also boiled in pans, placed in sheds near the wells from which it is drawn. Each shed is usually named from some animal or other thing which is pictured on a tablet nailed to it. The walls of these sheds are made either from baked earth or from wicker work covered with thick mud, although some may be made of stones or bricks. . . . Each shed is divided into three parts, in the first of which the firewood and straw are placed; in the middle room, separated from the first room by a partition, is the fireplace on which is placed the caldron. To the right of the caldron is a tub, into which is emptied the brine brought from the shed by the porters; to the left is a bench, on which there is room to lay thirty pieces of salt. . . . They construct the greater part of the fireplace of rock-salt and of clay mixed with salt and moistened with brine, for such walls are greatly hardened by the fire. These fireplaces are made eight and a half feet long, seven and three quarters feet wide, and, if wood is burned in them, nearly four feet high; but if straw is burned in them, they are six feet high. An iron rod, about four feet long, is engaged in a hole in an iron foot, which stands on the base of the middle of the furnace mouth. This mouth is three feet in width, and has a door which opens inward; through it they throw in the straw. The caldrons are rectangular, eight feet long, seven feet wide, and half a foot high, and are made of sheets of iron or lead, three feet long and of the same width, all but two digits. These plates are not very thick, so that the water is heated more quickly by the fire, and is boiled away rapidly. The more salty the water is, the sooner it is condensed into salt.'

The fine detail given in this work, both in the text and the illustrations, is astonishing, and sometimes amusing. In a later engraving of the above series, the salt manufacturer is shown in the background filling baskets with salt from a caldron, while in the foreground a woman is seated at lunch: the key below the picture contains the laconic and matter-of-fact explanation, ' D—Master, I—Baskets, F—Wife, P—Tankard which contains beer '. German and Italian translations of *De Re Metallica* were issued soon after the appearance of the original Latin edition; the only English translation is that of H. C. Hoover (later President Hoover) and Mrs L. H. Hoover, published in London in 1912, with reproductions of all the original illustrations and many lengthy and valuable annotations.[41]

It is of general alchemical interest that the title-page of *De Re Metallica* is ornamented with a caduceus, which the author associates with the divining rod and other potent rods. ' The Ancients,' he remarks,[42] ' by means of the divining rod, not only procured those things necessary for a livelihood or for luxury, but they were also able to alter the forms of things by it; as when the magicians changed the rods of the Egyptians into serpents, as the writings of the Hebrews relate; and as in Homer, Minerva with a divining rod turned the aged Ulysses suddenly into a youth, and then restored him back again to old age; Circe also changed Ulysses' companions into beasts, but afterward gave them back again their human form; moreover by his rod, which was called " Caduceus ", Mercury gave sleep to watchmen and awoke slumberers.'

As a rule Agricola maintains a high standard of discrimination and a strong sense of practical value in *De Re Metallica*; but in other works he succumbs occasionally to the credulity of his age. Thus, in 1549, he published at Basel a little book entitled *De Animantibus Subterraneis*. This deals with animals displaying subterraneous proclivities, and contains an unsuspecting reference to those hardy flies of Aristotle which were said to breed and flourish only in smelting furnaces! The same quaint publication contains an account of the fabled kobolds, or underground gnomes, of the mines. The beneficent variety of these beings evidently had much in common with the pixies of south-western England. According to Agricola,[43] ' They appear to laugh with glee and pretend to do much, but really do nothing. They are called little miners, because of their dwarfish stature, which is about two feet. They are venerable looking and are clothed like miners in a filleted garment with a leather apron about their loins. . . . Sometimes they throw pebbles at the workmen, but they rarely injure them unless the workmen first ridicule or curse them. . . . Because they generally

appear benign to men, the Germans call them *guteli*.'
Some of the figures of workmen in the illustrations of
De Re Metallica conform, curiously enough, to the above
description of these imaginary sprites of the mine.

Agricola's clear and practical treatment of his special
field of chemistry paved the way for Libavius' *Alchymia*,
which appeared in 1595. This celebrated work was a
general treatise on chemistry: expressed in an intelligible
and systematic form, it emphasised the practical aspect
of the subject. In these respects it followed closely the
model which Agricola had set up in *De Re Metallica*,
thirty-nine years earlier. Libavius' *Alchymia* was the
first work which has a claim to be considered as a text-
book of chemistry (p. 213). In general, writers of the
type of Brunschwick, Agricola, and Libavius made the road
along which later generations were able to pass from the
cribbed and cabined confines of alchemy into the spacious
domains of chemistry.

§ 9. ALCHEMICAL LITERATURE OF THE SEVENTEENTH CENTURY

Of all periods, the seventeenth century is the richest
in alchemical writings. Although it can now be seen
that alchemy was then on the wane, this century produced
a surprising efflorescence of treatises expounding and
defending alchemical doctrines, detailing marvellous
transmutations, and emphasising the allegorical, mystical,
and spiritual aspects of alchemy. The publications of
this last type are particularly characteristic of the declin-
ing days of alchemy; they present features of particular
interest, a number of which are considered in some detail
in Chapters III to VII.

The century opened with the picturesque story of
Alexander Seton, the so-called ' Cosmopolite ', and author
of *Novum Lumen Chymicum* (p. 47), who, travelling from
Scotland to the Continent, quickly achieved an unsurpassed
reputation as a successful adept. An imprudent demon-

stration of his alleged transmutative process led to his being immured in a dungeon and tortured by the Elector of Saxony, whose cupidity he had aroused; and although he succeeded in escaping in a dramatic manner, this 'chief martyr of Alchemy' died in 1604 as a result of the inhuman treatment he had received.[44]

The beginning of the seventeenth century was marked also by the appearance of a remarkable series of treatises ascribed to Basil Valentine (Chapter V), whom alchemical tradition assigned to the fifteenth century, before the advent of Paracelsus. There is little doubt, however, that the writings appearing under this name date from about the opening of the seventeenth century; they are racy of the alchemical soil of that age, and may be likened in many respects to other alchemical publications which undoubtedly belong to the first half of the seventeenth century.

Among such publications, the *Amphitheatrum Sapientiae Æternae* ('Amphitheatre of eternal wisdom') of Heinrich Khunrath, printed at Hanau in 1609, is the most remarkable work of this devotee of theosophy and the wider occultism. Khunrath was born at Leipzig in 1560 and graduated in medicine at Basel in 1588. His alchemy is spiritual rather than material. The curious cabbalistic plates of the *Amphitheatrum* depict the Emerald Table, the oratory-laboratory of Khunrath (Plate 52), the Alchemical Citadel, and other mystical and magical subjects, including Khunrath surrounded by his enemies—this astonishing engraving which is a veritable Callot in anticipation, as Caillet remarks, quoting from Stanislas de Guaita. A smaller illustration in this work, showing a bespectacled owl (Plate 10) holding two torches and flanked by a pair of lighted candles, carries the dark saying, 'What good are torches, light, or spectacles, to folk who won't see!'—

> Was helffen Fackeln/ Liecht oder Brillen/
> Wann die Leut nicht sehen wollen.

In another place Khunrath relapses from Latin into

German, in order to exclaim impatiently: ' He who sets out to make wise men of fools will be kept very busy '. Not being French, Khunrath was perhaps unacquainted with the contemporary saying:

Le Monde est plein de Fous, et qui n'en veut point voir,
Doit demeurer tout seul, et casser son miroir.*

Besides being intolerant, Khunrath was unbalanced. His writings consist largely of a fevered sequence of mystical pronouncements and adjurations, interlarded with bizarre exclamations in various tongues. Within half a dozen lines [45] he apostrophises the blessed Viridity, the universal principle of germination; urges the Theosophist to contemplate the Viridity, which is Ruah Elohim, the Spirit of the Lord; conjures the Cabbalist to regard the virid Line of the whirling universe; beseeches the Magist to meditate upon Nature; and exhorts the Physico-Chemist to brood over the green Lion and the mysteries of the Quintessence!

Waite [46] states that ' this German alchemist, who is claimed as a hierophant of the psychic side of the *magnum opus*, and who was undoubtedly aware of the larger issues of the Hermetic theorems, must be classed as a follower of Paracelsus '. The Alchemical Citadel is represented in a large double plate as a fortress guarded by a moat and surrounded by a maze of twenty-one sections. Of these, only one lends access across the moat; the other twenty carry inscriptions showing the false activities of the worldly-minded alchemists—for example, the attempt to transmute ordinary mercury. The successful adept must be endowed with a knowledge of the material of the

---

* For which the following free translation may be suggested:
' The world is full of fools, and he whose eye shuns asses,
Had better live alone, and break his looking-glasses.'
Besides being used by Khunrath, the device shown in Plate 10 appeared in a modified form, with a representation of the sun above the owl's head, in a curious little work published at Berlin in 1771 under the title *Eine grosse Herzstärkung für die Chymisten; nebst einer Dose voll gutes Niesepulver, für die unkundigen Widersprecher der Verwandlungskunst der Metalle.* The accompanying legend runs
' Du siehst beÿ Fackeln, Brillen, Licht
Und beÿ der hellen Sonne nicht.'

Great Work; also with faith, silence, purity of heart, and prayerfulness. After passing through the gate surmounted by the hieroglyph of philosophic mercury, he traverses the seven angles of the citadel, representing the chief operations of the Great Work (p. 136), such as calcination, dissolution, purification, introduction into the sealed Vase of Hermes, transference of the Vase to the Athanor, coagulation, putrefaction, ceration, multiplication, and projection; and even upon reaching the *Petra Philosophalis*, or Philosopher's Stone, he finds that it is held in custody by a formidable dragon.[47]

In a striking symbolical representation of the analogy of the alchemical microcosm to the macrocosm, found in the *Basilica Philosophica* (' Portico of Philosophy ') of Johann Daniel Mylius (Frankfurt, 1618), a still broader pictorial interpretation is given of the place of alchemy in the universe. Here, proceeding from ' what is above ' to ' that which is below ' (p. 54), we obtain a comprehensive view of ' the miracles of one thing '. First comes the celestial world with the Tetragrammaton, or Name of the Lord, surrounded by angels. Next appears the planetary and zodiacal world. This is followed by the terrestrial world, in the centre of which appears the Garden of Alchemy fully planted with the trees of the seven metals, the three principles, tartar, nitre, orpiment, sal ammoniac, verdigris, and certain others. Below, on the left, is man, carrying the symbol of Sol, or sophic sulphur, the masculine principle; on the right, woman, with the symbol of Luna, or sophic mercury, the feminine principle; the woman stands in the Hermetic stream (p. 63),* and holds in her right hand a bunch of grapes, significant of fertility. Both these principles take part in the alchemical operation, and are linked to the macrocosm by chains—' bound by gold chains about the feet of God '.[48] The creation of the world, the generation of

---

* The horned figure facing Luna is suggestive of the metamorphosis of Actaeon into a stag, at the sight of Diana (Luna) bathing (in the Hermetic stream).

metals, and the reproduction of man are held to be fundamentally similar. At the bottom of the picture, the phoenix (left) is associated with two globes symbolising fire and air, and the eagle (right) with two other globes representing water and earth. The double-bodied lion is a symbol of the blending of the ' two sulphurs ' (p. 221), and the surmounting figure is possibly the Elder (p. 156); the representations of fire behind one body (left) and the Hermetic stream behind the other (right) are again indicative of sophic sulphur and sophic mercury.

This illustration is reproduced as Figura IV of *Janitor Pansophus* (' The All-Wise Doorkeeper '), at the end of the *Musaeum Hermeticum* (Frankfurt, 1678), in the form of a fine engraving (Plate 11) measuring about $13\frac{1}{2}$ by $10\frac{1}{2}$ inches. It provides a striking example of the widespread and misleading use of analogy in mediaeval science, and is suggestive of the ever-recurring analogy between the body and life of man (the microcosm) and the matter and forces of the outside universe (the macrocosm). Mediaeval science abounds in such evidences of man's conviction of his unity with Nature, and of his determination to see his own image in Nature and Nature's image in himself : ' All that is without thee also is within ' (p. 73).

Part of the symbolical scheme of Mylius appears to find a reflection in a much simpler drawing occurring as a frontispiece to an alchemical manuscript of the eighteenth century untitled and written in German, in the St. Andrews collection (Plate 12). In the firmament above are shown the seven heavenly bodies radiating their influence upon the earth beneath, thereby, according to alchemical doctrine, favouring the generation of the corresponding metals; converging also upon the central crowned globe are emanations proceeding from a crowned eagle and lion, representing Wind and Earth of Hermes' fourth precept, or sulphur and mercury, or the volatile and fixed principles. The crown symbolises the Philosopher's Stone, which was sometimes called the Great King. The mingled waters issuing from the mouths of three crowned lions, carrying the symbols of mercury, salt, and sulphur, give rise in the bowels of the earth to the seven metals, together with antimony and vitriol. The animation of the whole work by the breath of the four elements is perhaps allusive to the Spirit of God moving upon the face of the waters in the creation of the world; for it has been held [49] that ' the first chapter of Genesis is the greatest page in alchemy '.

Other works of this period, in particular *Atalanta Fugiens*, published in 1618 by Michael Maier (p. 236), are characterised by a crude anthropomorphism linked with classical mythology and phallic ideas. To Michael Maier also must perhaps be ascribed a curious geometrical representation (Fig. 8) of the place of man and alchemy in the cosmos; this figure appeared as an interpolation in the Latin translation [50] of Norton's *Ordinall*, published under Maier's supervision in 1618 (p. 174), and it was republished in the *Musaeum Hermeticum* seven years later. In the most exalted zone are inscriptions denoting (I) The Archetypal World, God, Jehovah, Infinite Good. Then follows a zone of (II) Angels, (III) Ether, (IIII) Elements, and Finite Good; the four associated triangles, with Man (Homo) at the centre, are inscribed Heaven, Earth, Air, Water. The names at the numbered angles are (1) Mercury, (2) Salt, (3) Sulphur, (4) Stars, Birds, Fishes, (5) Metals, Planets, Stones, (6) Beasts, Meteors, Angels. The ' content ' of each small triangle is interesting. The nethermost zone (V) consists of the Infernal Regions, with the progressively portentous inscriptions— Fire, Winds, The Void, Darkness, Abyss, Chaos, Evil, Satan.

§ 10. Symbols, Emblems, and Cryptic Expression

The *Summa Perfectionis* is an alchemical treatise distinguished by the clarity of its expression and the practical utility of much of its matter. These characteristics are displayed in a still higher degree by purely utilitarian

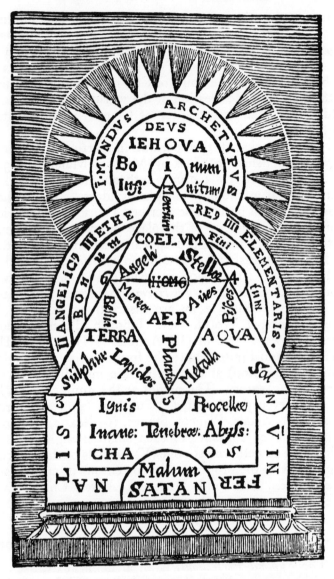

Fig. 8.—Man, Alchemy, and the Cosmos.
From *Bibliotheca Chemica Curiosa*, Manget, 1702.

publications, such as Brunschwick's *Buch zu Distillieren* and Agricola's *De Re Metallica*. At the other extreme stand the purely mystical works, of which examples have been considered in *Splendor Solis* and Khunrath's *Amphitheatrum*. On the whole, alchemical literature leans towards the latter type. The outstanding features of the great mass of the writings are their symbolism and mysticism; their addiction to allegory and cryptic expression; and their general atmosphere of prolix nebulosity.

Trismosin [51] wrote that ' the noble Alchemy is to be esteemed as the gift of God; for it is hidden mostly in manifold proverbs, figurative sayings and parables of the old Sages '. Some writers took an open pride in their use of cryptic language: ' all haile to the noble Companie of true Students in holy Alchimie', exclaims the author of *The Hunting of the Green Lyon*,[52] ' whose noble practise doth hem teach to vaile their secrets with mistie speach '. Many of them wrote as men possessed of esoteric knowledge, or weighty secrets, which were to be reserved for an elect brotherhood. These held that if their esoteric knowledge were revealed in all its simplicity, women and even children would be able to exploit the wonders of the 'Divine Art'. The dullest clodhopper would turn away from his plough in order to emulate Jason by cultivating the more alluring soil of the Sages (p. 94). Riches and longevity would become universal. . . . But what a mistake, they observed, to feed donkeys on lettuces, when thistles are sufficient! [53]   In ancient Egypt a knowledge of chemical secrets was the guarded prerogative of priests and kings, and it was perhaps the mantle of Thoth descending upon the ' sons of Hermes ' which led them to enmesh their ' precious pearls' of alchemy in a web of enigma, cryptogram, and obscure expression, far surpassing in unintelligibility their patron's famous Smaragdine Table.

The symbolism of alchemy was a carefully tended heritage derived also from a remote past. Symbols were used, before the development of adequate spoken or written languages, for the purpose of conveying abstract ideas. The early mythological symbols formed the beginnings of an extensive symbolic system which ultimately embraced not only the numerous attributes of Divinity but also the still more numerous manifestations of Nature. We have already seen (p. 20) that in ancient China the Five Elements were associated with the planets, and the Two Contraries with the sun and moon. The fundamental symbols applied in Western alchemy were an outcome of the similar Chaldean practice of associating these seven familiar heavenly bodies with the seven metals, and with human organs and destinies. In this symbolism, Sol (gold) was represented by a circle, and Luna (silver) by a crescent. The five ' base ' metals carried the planetary signs of Mercury (quicksilver), Venus (copper), Mars (iron), Jupiter (tin), and Saturn (lead).* The symbols were often ornamented, as shown, for example, in Fig. 17. They were sometimes accompanied by representations of the corresponding Olympian deities (Plate 13).

These signs have formed the subject of much comment and speculation. The derivations of the first two are obvious: gold was dedicated to the Sun and silver to the Moon, the Father God and Mother God of the ancient religions with male and female tenets. The lunar symbol appears also in such guises as the crescent or cup of various goddesses of rain and fertility, and the horseshoe of folklore. The other signs have been variously interpreted as the caduceus of Mercury; the looking-glass of Venus; the spear and shield of Mars; the thunderbolt of Jove, the Arabic 4 for the fourth planet, or the Greek initial for Zeus; the scythe of Saturn, or the initial of Kronos. Tin, on account of its brightness and the crackling ' cry ' which it emits when bent, invited recognition as the thunderbolt of Jupiter.

Some authorities assign a phallic origin to the symbols

* This association, although usual, was not invariable: in ancient Persia, for example, tin was coupled with Venus and iron with Mercury.

PLATE 19

Melchior Cibinensis at Mass.

From *Viridarium Chymicum*, Stolcius, 1624 (reprinted from *Symbola Aureae Mensae Duodecim Nationum*, Maier, 1617). (See p. 114.)

PLATE 20

Wei Po-yang, his dog, and his disciple, Yü, with a furnace (*ting*).

Reproduced from a Japanese imprint, about 500 years old, in a copy of the *Lieh Hsien Ch'üan chuan* ('Complete Biographies of the Immortals').

PLATE 21

Ubiquity of the Philosopher's Stone.

From *Atalanta Fugiens*, Maier, 1618. (See pp. 130, 244.)

PLATE 22

(i) Sol (Sulphur), Luna (Mercury), and the Hermetic Stream.

From *Elementa Chemiae*, Barchusen, 1718. This design symbolises the remote and proximate materials of the Stone. (See p. 133.)

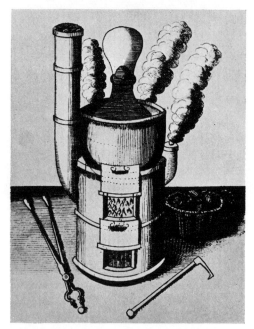

(ii) Vase of Hermes and Athanor.

A representation of the sealed vessel of Hermes and the furnace used in the Great Work. From the same. (See p. 152.)

of Mars and Venus: the latter, in particular, is closely
akin to the ' handled cross '—the Egyptian *'ankh* or *crux
ansata* (p. 109)—the symbol of the Sun's gift of life to the
world (p. 34). Gold was probably assigned to the Sun
because of its colour and perfection, silver to the Moon
for similar reasons, and quicksilver to Mercury because of
the quick motion of this, the only liquid, metal. Iron, the
metal of war, was associated with Mars; and dull, heavy
lead with slow-moving Saturn. The symbol of lead is
often displayed on a darkened field (Plate 18 (ii)).
Glauber estimated the degrees of perfection of the metals
from the number of sides of an enclosing square which
made contact with their symbols (p. 182).

It is interesting that these seven metals, through their
corresponding heavenly bodies and gods, are associated
with the seven days of the week: the Sun and Moon
with Sunday and Monday; Mars with Tuesday (*Tiwes
daeg*, from the Anglo-Saxon *Tiw*, Mars); Mercury with
Wednesday (after Woden), in French, *mercredi* (*mercurii
dies*); Jupiter with Thursday (after Thor), in French,
jeudi (*jovis dies*); Venus with Friday (after Frigg, the wife
of Woden), in French, *vendredi* (*veneris dies*); and Saturn
with Saturday.[54]

This ancient association has persisted to the present
day in such names as ' lunar caustic ' (silver nitrate) and
' extract of Saturn ' (lead acetate solution). It figures
prominently in mediaeval literature, the best-known reference being in Chaucer's *Canones Yeomans Tale*:[55]

> The bodies seven, eek, lo heer anon.
> Sol gold is, and Luna silver we declare;
> Mars yron, Mercurie is quyksilver;
> Saturnus leed, and Jubitur is tyn,
> And Venus coper, by my fathers kyn.

John Gower[56] versified the relationships in closely similar
terms, except that he assigned ' the Brasse ' to Jupiter.

Gold and silver (Sol and Luna), iron and copper (Mars
and Venus), and lead and quicksilver (Saturn and Mercury)

FIG. 9.—Alchemical Hieroglyphics.

From the *Last Will and Testament* of Basil Valentine, London, 1671.

are frequently treated in alchemy as pairs of opposites. Gold and silver commonly, and iron and copper occasionally, represent the male and female principles in exoteric alchemy; also lead, the dull and heavy metal, associated with slow-moving Saturn (Kronos) and sometimes symbolised by a wooden-legged man with a scythe (Plate 37 (i)), is often regarded as the antithesis of quicksilver, associated with Mercury (Hermes). Thus, one of the Figures of Abraham the Jew (p. 63) symbolises the 'fixing' of quicksilver by means of lead.

Owing perhaps to their abhorrence of Greek paganism, the Arabs were sparing in their use of the considerable variety of Greek symbols, and it was not until alchemy had become established in western Europe that its practitioners developed the comprehensive code which permeates mediaeval alchemical writings. In designing symbols for the elements and principles (pp. 11 and 208), simple geometrical figures were introduced. Gradually, an imposing array of hieroglyphics came into use; but, unfortunately, and probably of set purpose, there was no uniformity in their application, as will be evident from a glance at a list given in Basil Valentine's *Last Will and Testament* [57] (Fig. 9): gold, for example, was represented at one time or another in more than sixty ways. To add to the confusion, anagrams, acrostics, and other enigmas were introduced, and various secret alphabets and ciphers came to be used by alchemists; in some of these, letters and numerals were represented by alchemical and astrological signs. An additional barrier was erected in the shape of an extensive structure of pictorial symbolism and allegorical expression. Ideas, processes, even pieces of apparatus, were represented by birds, animals, mythological figures, geometrical designs, and other emblems born of a riotous, extravagant, and superstitious imagination.

By such means, the results of alchemical experience were effectively segregated. A free and continuous interchange of knowledge and experimental observation is the life-blood of modern science. The abnegation of this principle by the alchemists was largely responsible for the stagnation of alchemy. It may be observed, however, that alchemy came into being more than a thousand years before the invention of printing (Fig. 3). Moreover, the alchemists attached little or no importance to securing ' priority of publication ', or, for that matter, publication at all. The ' puffer ' was jealous of his competitors in the quest for gold; and the true adept considered himself a chosen guardian of the secrets of the 'Divine Art'. To the adept, his work was a religious activity rather than a stepping-stone to ' great dignity and fame ' (p. 178). Thus, masters of the ' holy Art ' like Thomas Norton recorded their work only in cautiously worded manuscripts, for which they claimed no credit. ' *For they being lovers of* Wisdome *more than* Worldly Wealth, *drove at* higher *and more* Excellent Operations: *And certainly* He *to whom the whole* Course *of* Nature *lyes open, rejoyceth not so much that he can make* Gold *and* Silver . . . *as that he sees the* Heavens *open, the* Angells *of* God *Ascending and Descending, and that his own Name is fairely written in the* Book of life.' An appreciation of this fundamental point of view, expressed so quaintly by Elias Ashmole,[58] is essential in assessing the vexed and many-sided problem of alchemical anonymity.

The emblems of alchemy are often very quaint and picturesque, especially as many of them conform to a system of colours suggestive of the principles of heraldry. They include the red king (sophic sulphur, or gold), the white queen (sophic mercury, or silver), the ascending dove or swan (a white sublimate), the grey wolf (antimony), the black crow (black or putrefying matter), the toad (earthy matter), winged and wingless lions (mercury and sulphur), and endless others. Philalethes [59] in an exuberant mood let loose the following deluge of synonyms for mercury: ' Mercury is our doorkeeper, our balm, our honey, oil, urine, may-dew, mother, egg, secret furnace, oven, true fire, venomous Dragon, Theriac, ardent wine, Green Lion, Bird of Hermes, Goose of Hermogenes, two-edged sword in the hand of the Cherub that guards the Tree of Life, etc. etc.; it is our true, secret vessel, and the Garden of the Sages, in which our Sun rises and sets. It is our . . .' and he inverts a second nomenclatory cornucopia upon the head of the unfortunate neophyte. As Liebig [60] remarks, the alchemists ' propounded in an unintelligible language that which, in their own minds, was only the faint dawn of an idea '.

The unrestrained use by the alchemists of symbols, emblems, allegory, and other forms of cryptic and mystical expression, led to the production of a unique literature, laden with a mass of enigma, metaphoric expression, and jargon which defies description. Ben Jonson, who displays in *The Alchemist* an unrivalled mastery of the vocabulary and imagery of alchemy, savours this atmosphere when he talks [61] of ' the pale citron, the green lion, the crow, the peacock's tail, the plumèd swan . . . the flower, the *Sanguis Agni* . . . your oil of height, your tree of life, your blood . . . your Lato, Azoch, Zernich, Chibrit, Heautarit. . . .'

> The bulls, our furnace,
> Still breathing fire: our argent-vive, the dragon:
> The dragon's teeth, mercury sublimate,
> That keeps the whiteness, hardness, and the biting:
> And they are gathered into Jason's helm,
> (The alembic), and then sowed in Mars his field,
> And thence sublimed so often, till they're fixed.
> Both this, th' Hesperian garden, Cadmus' story,
> Jove's shower, the boon of Midas, Argus' eyes,
> Boccace his Demogorgon, thousands more,
> All abstract riddles of our stone.

These typical selections from the alchemical vocabulary may be supplemented by an example of another sort of cryptic expression of a numerical kind (p. 248), taken from Elias Ashmole's *Fragments*: [62]

> Abowte 653. I dare be bold,
> This *Chyld* shall put on a Crown of Gold;
> Or at 656. at the moste,
> *This Chyld* shall rule the roste.

The ' child ' is here the Philosopher's Stone, which was sometimes represented as a new-born babe.

§ 11. HYLOZOISTIC CONCEPTIONS IN ALCHEMY

The symbols, emblems, and cryptic language of alchemical literature are inseparable from the main and subsidiary beliefs of alchemy, and some of the latter must be outlined at this point. Since alchemical theory assigned a body, soul, and spirit to metals and other inanimate materials (p. 27), it is not surprising to find that the analogy with man—and with living things in general —was carried further. Such hylozoistic conceptions may be traced back to Greece and the older civilisations (p. 12). In particular, the alchemists sought to apply to the inanimate world the principles of growth and reproduction which impressed them so much in the world of plants and animals. Thus, they were led to imagine that metals and ores grow in the earth, and further, that during the process of growth a ' base ' or imperfect metal might change slowly into the perfection of gold. However preposterous such ideas may appear, it is well to recall that in the present century English peasants have held firmly to the belief that stones are replenished in tilled fields by fresh growth. Moreover, even the sophisticated chemist pays tribute to these old alchemical theories and notions when he makes use of such expressions as ' golden sunlight ', ' the silvery moon ', ' native gold ', ' hewn from the living rock ', ' noble metals ', ' the fury of the elements ', and ' diamond of the first water '.

The conception of the growth of metals called for a metallic seed, and the search for the seed of gold was inextricably intertwined with the quest of the Philosopher's Stone: the seed of gold, or ' chrysosperm ', was said to be ' lodged in all metals '. The perfect seed would produce gold; imperfect seed would lead to imperfect, or aborted, metals. ' Sow your gold in white earth made of leaves ', runs the injunction attached to an alchemical drawing of 1618, illustrating this conception (Plate 14). In extending the supposed analogy, the mistaken idea that a plant seed must putrefy, or die, before it can germinate (p. 202) was transferred to the supposed seed of metals: thus, alchemical literature abounds in references to the ' putrefaction ', ' mortification ', and ' killing ' of metals or their seeds, followed by ' revivification ' or ' resurrection '.   It was imagined also that art might improve upon Nature and accelerate the supposed changes.   Basil Valentine[63] held that animals, vegetables, minerals, and metals all ' have their original seed from God . . . note farther, if any of these Metalline and Mineral kinds shall be brought to a farther propagation and augmentation, it must be reduced to its first seed and *prima materia* '.

Even here the alchemists could adduce experimental evidence in support of their theoretical scheme.   ' They would take a metal, say lead, and calcine it in the air. They watched it lose its well-known appearance and change into a powdery kind of cinder.   Assuming, as they did, that the metal had a life of its own, what more natural than to say that it had died?   It was in the condition which they imagined a seed would be that had died in the ground.   They then reheated this cinder in a crucible along with some grains of wheat.   They watched the metal taking on again its wonted characteristics and resuming its original state.   What was more natural than to suppose that the life in the grain had brought about a " resurrection " of the metal? '[64]

§ 12. ALCHEMY AND ASTROLOGY

In his Epistle to *The Compound of Alchymie*[65] Ripley writes:

We have an Heaven yncorruptible of the Quintessence,
Ornate with Elements, Signes, Planetts, and Starrs bright,
Which moisteth our Erthe by Suttile influence:
And owt thereof a Secrete Sulphure hid from sight,
It fetteth by vertue of his attractive might;
Like as the Bee fetcheth Hony out of the Flowre.

This quotation illustrates a further factor in alchemical theory, namely, the supposition that astral influences entered into the processes of generation and growth: the sun favoured the production of gold, the moon of silver, and so forth. An astral origin of the metals is indicated by Thomas Norton [66] in the following words:

> For cause efficient of Mettalls finde ye shall
> Only to be the vertue Minerall,
> Which in everie Erth is not found,
> But in certaine places of eligible ground;
> Into which places the Heavenly Spheare,
> Sendeth his beames directly everie yeare.
> And as the matters there disposed be
> Such Mettalls thereof formed shall you see.

Pictorial expressions of such ideas are often found in alchemical works, as, for example, in the *Musaeum Hermeticum* (Plate 31).

'All corporeal things originate in and are maintained and exist of the Earth, according to Time and Influence of the Stars and Planets,' wrote Trismosin [67] in similar vein. This belief, derived from ancient Chaldea, is associated in turn with a wide field of folk-lore dealing with the growing and gathering of herbs and crops under the proper astrological auspices. 'We may ordinarily observe', wrote Elias Ashmole [68] in 1652, 'how *poorely* and *slowly* the *Seeds* of *Plants* grow up, nay many times *languish* and *degenerate* into an unkindly *Quality* and *Tast*, if sowne in the *Waine* of the Moone, and the Reason is because the *Moysture* and *Sapp* that should feed them is exceedingly diminished; yet it is the fittest tyme for cutting down *Timber*, or what else we would preserve from decaying.'

The alchemists attached great importance to the doctrine of signatures, a quaint conception which received universal credence throughout the alchemical era. In an interesting disquisition upon this species of cipher, by means of which Nature was held to convey valuable information to man, Ashmole [69] wrote as follows: 'As for the use of such *Characters*, *Letters*, *Words*, *Figures*, &c. Formed or Insculped upon any *Matter* we make use of, we are led to it by the president of *Nature*, who *Stampes* most notable and marvelous *Figures* upon *Plants*, *Roots*, *Seeds*, *Fruits*, nay even upon rude *Stones*, *Flints*, and other inferior *Bodies*.

' Nor are these remarkable *Signatures* made and described by Chaunce, (for there is a certaine *Providence* which leades on all things to their end, and which makes nothing but to some purpose,) but are the *Characters and Figures* of those *Starrs*, by whom they are principally governed, and with these particular *Stamps*, have also peculiar and different *vertues* bestowed upon them. What *Artists* therefore doe in point of *Character*, is onely to pursue the Track, that is beaten out by *Nature*; And by how much the more the *Matter* whereupon such *Impressions* are made, is sutable to the *Qualities* of those *Starrs* whose *Characters* it is signed with: By so much more apt and inclineable it will be to receive those *vertues* that shall impower it to produce an *Effect*, in things whereunto it's applyed. . . . And this is not strange if we reflect upon the Vulgar experiments of the *Loadestone*, who communicating its vertue to a peece of *Iron* (a thing made fit by *Nature* to attract and reteine) that *Piece* thereby becomes of strength to communicate this vertue to a *third*. But if we should consider the *Operations* of this *Magnet* throughly (which proceeds onely from a *Naturall Principle*) there is no other *Mystery*, *Celestiall*, *Elementall*, or *Earthly*, which can be too hard, for our *Beliefe*.'

There is a certain plant which possessed a peculiar significance in alchemy because of its imagined 'impregnation with a celestial vitality ' derived from the moon. Of this so-called Lunary, an anonymous poet quoted by Ashmole [70] writes

> Her ys an Erbe men calls *Lunayrie*,
> I-blesset mowte hys maker bee.
> *Asterion* he ys I-callet alle so,
> And other namys many and mo;

> He ys an Erbe of grete myght,
> Of Sol the Sunn he taketh hys lyght . . .
> The Sune by day, the Mone by nyght,
> That maketh hym both fayre and bryght . . .
> The Rote ys blacke, the Stalke ys red;
> The wyche schall ther never be dede . . .
> Hys Flowrys schynith, fayre and cler,
> In alle the Worlde thaye have non pere . . .
> With many a vertu both fayre and cler,
> As ther ben dayes in alle the yere,
> Fro fallyng Ewel and alle Sekeneys,
> From Sorowe he brengyth man to Bles.

Of this herb (*Botrychium Lunaria*), Gerard[71] says: ' Small
Moone-wort is singular to heale greene and fresh wounds.
It hath beene used among the Alchymists and witches to
doe wonders withall, who say, that it will loose lockes, and
make them to fall from the feet of horses that grase where
it doth grow, and hath beene called of them *Martagon*
[p. 178], whereas in truth they are all but drowsie dreames
and illusions; but it is singular for wounds as aforesaid '.
In the light of the doctrine of signatures, the alchemists
attached great importance to the likeness of the crescent
moon which they discerned in the leaves of this plant: *

> The Levvis ben rownd, as a Nowbel son,
> And wexsyth and wanyth as the Mon:
> In the meddes a marke the brede of a peni,
> Lo thys is lyke to owre sweght Lunayre.

Similarly, in *Bloomefields Blossoms*[72] it is said that ' The
Moone that is called the lesser *Lunary*, Wife unto *Phoebus*,
shining by Night, To others gives her Garments through
her hearb *Lunary* '. According to some alchemical
writers, Lunary had a root of metallic earth; moreover, it
was said to put forth citrine flowers after three days, and
to change into silver when put into mercury.[73] Sophic
mercury was sometimes called ' essence of Lunary ', or
' moon-spittle '. This plant provides an example of the
many links between alchemy and those fragrant reposi-

---

* See the reproduction of Ashmole's illustration on p. xxiv.

tories of the lore of plants and plant-lovers—the herbals
of the sixteenth and seventeenth centuries.

The familiar alchemical emblem of the ' tree of life ' of
the seven metals, or the tree of ' universal matter ', illustrates the idea of astrological influence in a similar way,
the fruit of this tree consisting of symbols representing
the metals and their tutelary planets (Fig. 17, p. 269).
Sun-trees and moon-trees are often depicted (Plate
16), and occasionally the metals and planets are portrayed as flowers (p. 210), the sun and moon being rose
and lily respectively. Sometimes, also, birds of various
colours are shown in an oak or other tree. Often a Hermetic stream, or spring, representing the elixir of life, water
of the wise, or spirit of the metals, flows from its roots.

The foregoing examples illustrate one phase of an
ancient, comprehensive, and far-reaching system, in which
alchemy and astrology were blended. It has already been
pointed out (p. 88) that the Chaldeans coupled the seven
prominent heavenly bodies with the seven known metals,
and that in ancient China these bodies were linked with
the Five Elements and the Two Contraries. The Chinese
alchemists attributed considerable importance to astrological considerations: thus, Wei Po-yang [74] assigns the
Blue Dragon to the *Fang* constellation, of spring; the
White Tiger to *Mao*, of autumn; and the Scarlet Bird to
Chang, of winter.

Alchemical literature abounds in reflections and de-
velopments of these primitive ideas. To select one of
the clearest and most definite allusions, Norton (p. 174)
writes as follows in his *Ordinall of Alchimy* (1477): [75]

> The *fift Concord* is knowne well of *Clerks*,
> Betweene the *Sphere of Heaven* and our *Suttill Werks*
> Nothing in Erth hath more Simplicitie,
> Than th' elements of our *Stone* woll be,
> Wherefore thei being in warke of Generation,
> Have most Obedience to Constellation . . .
> For the *White warke* make fortunate the *Moone*,
> For the Lord of the *Fourth house* likewise be it done . . .

Save all them well from greate impediments,
As it is in Picture, or like the same intents.

Here, in the words of Elias Ashmole,[76] ' our *Author* refers
to the *Rules* of *Astrologie* for *Electing* a time wherein to
begin the *Philosophicall* worke, and that plainly appeares
by the following lines, in which he chalkes out an *Election*
fitly relating to the Businesse. In the *operative* part of
this *Science* the *Rules* of *Astronomie* and *Astrologie* (as elsewhere I have said) are to be consulted with.' Ashmole
adds that in these ' Elections ' the ' calculatory part '
belongs to astronomy and the ' judiciary ' to astrology.
In words which are reminiscent of the eighth precept of
Hermes (p. 54) he says that ' since both inferiour and
superiour Causes concur to every effect, it followeth that
if the one be not considered as well as the other, this negligence will beget error '. Such ' elections ', indeed, are
held to make use of celestial influences just as a physician
makes use of the variety of herbs.

In reproducing Norton's astrological ' scheames '
(Plate 15), with their ' Signes and Figures and parts
aspectuall ', Ashmole makes certain technical criticisms of
the figures, but tempers his remarks with the observation
that in them ' the *Author* continues his *Vailes* and *Shadows*,
as in other parts of the *Mistery* '. Astrology, he observes
ponderously, ' is a profound Science: The depth this *Art*
lyes obscur'd in, is not to be reach't by every vulgar
Plumet that attempts to sound it. . . . There are in
Astrologie (I confesse) shallow *Brookes*, through which
young *Tyroes* may *wade*; but withall, there are deepe
Foards, over which even the *Gyants* themselves must *swim*.'

Two points of astrological significance deserve special
mention. The constellation of the Dragon, owing to its
situation at the pole of the zodiacal circle, was endowed
with a peculiar astrological potency. ' The Celestial
Dragon is placed over the Universe like a king upon a
throne ', wrote the Cabbalist in the *Book of Creation*.
This central symbol of the sky was an image of power and
energy, and perhaps also of time. The adoption of the

dragon as the symbol of mercury (p. 107) may have followed from the observation that the seasonal aspects of this constellation could be correlated with the movements of the planet Mercury. To the Moon also, the celestial symbol of the feminine principle, was attributed special astrological powers, because of tidal and physiological periodicities which were associated with its motion through the heavens.

§ 13. MASCULINE AND FEMININE PRINCIPLES

The idea of masculine and feminine principles entering into the generation of the Philosopher's Stone, and of metals and other inanimate materials, is widely disseminated in alchemical literature. We have already seen that this fundamental notion of the origin of all material things was held in common in ancient China, India, and Egypt (p. 20). It persisted throughout the era of alchemy. Thus, in the heyday of Western alchemy, Norton[77] refers in the *Ordinall of Alchimy* to ' the faire White Woman married to the Ruddy Man ', and Ripley[78] also writes of ' the *Red Man* and hys *Whyte Wyfe* '.

Flamel[79] states more explicitly: ' In this second operation, thou hast . . . two *natures* conjoyned and married together, the *Masculine* and the *Foeminine* . . . that is to say, they are made one onely body, which is the *Androgyne*, or *Hermaphrodite* of the *Ancients*, which they have also called otherwise, *the head of the Crow*, or *natures converted* '. In the engraving of Flamel's frescoes reproduced in Plate 7, this idea is represented pictorially by the figure numbered (3) and depicting a ' man and woman clothed in a gowne of Orange colour upon a field azure and blew '. In Flamel's ingenuous language: ' The *man* painted here doth expressly resemble *my selfe* to the naturall, as the *woman* doth lively figure *Perrenelle* . . . it needed but to represent a *male* and a *female* . . . but it pleased the *Painter* to put us there '.

Many cryptic expressions were used by the alchemists to denote man and wife: one of the quaintest was ' milk and cream ', the wife being the cream. Sometimes it was even held that of the four elements two were male and heavy, and two were female and light.

The King and Queen, or Sol and Luna, were the alchemical designations most commonly used for the masculine and feminine principles respectively. In exoteric alchemy, Sol and Luna denoted gold and silver; but in the esoteric literature and operations of the adepts these terms usually stood for sophic sulphur and sophic mercury respectively—or for ' our sulphur ' and ' our mercury ', as they were often called. The best source of sophic sulphur was commonly held to be gold; and thus such terms as sophic sulphur, gold of the philosophers, and seed of gold are to a large extent synonymous in alchemical literature. For this reason, both gold and sulphur—of the sophic kind—were identified by the adepts with the masculine principle and represented by the symbol of Sol, which had always been associated with ordinary gold. Again, the feminine principle, sophic mercury, was linked with silver, owing to an imagined relationship between quicksilver (argent-vive) and silver (argent): the exoteric symbol of Luna was therefore transferred by the adepts to denote sophic mercury. This dual application of the symbols of Sol and Luna has given rise to much confusion of thought, not only among the original alchemists, but also among their commentators.

In the operations of the Great Work, the union of masculine and feminine principles was associated with the process known as conjunction; the union was sometimes represented by hermaphroditic designs (Plates 16 and 60 (i)). Sulphur was said to bestow, and mercury to receive, the form assumed by the material resulting from their conjunction—just as wax takes and retains the impression of a seal. There were innumerable synonyms for each of these principles, many of which were used also to denote the Philosopher's Stone (p. 127). Sophic sulphur was called Sol, king, male, brother, Osiris, lion, toad, wingless dragon, natural fire, fixed principle, seal of Hermes, red mercury, incombustible oil, etc. Sophic mercury was disguised under such names as Luna, queen, female, sister, Isis (also Latona and other mythological names), serpent, eagle, winged dragon, humid radical (also celestial water, or rain of the sages), volatile principle, wax for Hermes' seal, white mercury, menstruum, bath of the king, etc.

Sometimes the masculine and feminine principles, or alternatively the two principles of the sulphur-mercury theory, were called the two sulphurs (Plate 57 (i)), sometimes also the two mercuries. The term 'mercury' is one of the chief stumbling-blocks of alchemical nomenclature, since it was used also for the ordinary metal (quicksilver or *argent-vive*), the proximate material of the Great Work (sophic mercury), and even for the Stone itself. A particularly intriguing association was that of the cat with Luna, or sophic mercury, because the pupil of a cat's eye was said to expand when the moon was waxing and to contract when it was waning. 'What's your mercury?' asks Subtle, in *The Alchemist*.[80] 'A very fugitive, he will be gone, sir', answers Face. 'How do you know him?' 'By his viscosity, his oleosity, and his suscitability.' Ben Jonson gives also a clear picture of sulphur and mercury as male and female principles:[81]

> Of that airy
> And oily water mercury is engendered;
> Sulphur o' the fat and earthy part; the one
> (Which is the last) supplying the place of male,
> The other of female in all metals.
> Some do believe hermaphrodeity,
> That both do act and suffer. But these two
> Make the rest ductile, malleable, extensive.
> And even in gold they are; for we do find
> Seeds of them by our fire, and gold in them;
> And can produce the species of each metal
> More perfect thence, than nature doth in earth.

PRELUDE TO CHEMISTRY

The Philosopher's Egg, also, played a great part in alchemical writings (p. 149) and operations. In the laboratory this was an oval glass vessel with a neck capable of being 'hermetically sealed'; it was also known as Hermes' Vase, or the Hermetic Vase. Figuratively, the egg was regarded by the alchemists, as by the ancient civilisations, as a symbol of creation; the Greeks approached still nearer to the alchemical point of view by envisaging it as a container of the four elements. A delineation (Plate 17) of the Philosopher's Egg, published in *Atalanta Fugiens* of 1618, carries an allusive reference to the Philosopher's Stone: 'There is a bird in the world, higher than all, the egg whereof to look for, be thy only care. A yellowish white surrounds the soft yolk; aim at this carefully, as is the custom, with the fiery sword; let Mars [iron] lend his aid to Vulcan [fire], and thence the chick arising will be conqueror of iron and fire.' Ben Jonson makes play with the same conception in *The Alchemist*: [82]

Surly. That you should hatch gold in a furnace, sir,
As they do eggs in Egypt! . . .
Subtle. Why I think that the greater miracle.
No eggs but differ from a chicken more
Than metals in themselves.
Surly.   That cannot be.
The egg's ordained by nature to that end,
And is a chicken *in potentiâ*.
Subtle. The same we say of lead, and other metals,
Which would be gold, if they had time.
Mammon.   And that
Our art doth further.
Subtle.   Aye, for 'twere absurd
To think that nature in the earth bred gold
Perfect i' the instant. Something went before.
There must be rémote matter.

An enigmatic design (Plate 18 (ii)) appearing in works attributed to Basil Valentine [83] illustrates some of the points under discussion. The planets shown in the firmament exercise their influence upon the combination

PLATE 23

(i) The Four Degrees of Heat.

From *Viridarium Chymicum*, Stolcius, 1624 (reprinted from
Philosophia Reformata, Mylius, 1622). (See pp. 144, 264.)

(ii) The Water of the Sages.

From the same. (See p. 148.)

PLATE 24

Ben Jonson (1573–1617).
Gerard Honthorst *pinxit*, Philip Audinet *sculpsit*.

PLATE 25

Collecting May-Dew for the Great Work.

From *Mutus Liber* in *Bibliotheca Chemica Curiosa*, Manget, 1702. (See p. 157.)

PLATE 26

Processes of the Great Work.

From an 18th-century French version of *Mutus Liber*. (See p. 158.)

THE LITERATURE OF ALCHEMY

of the masculine and feminine principles, represented by
the sun and moon.   In the chalice, the symbol of fruitfulness, may perhaps be discerned a likeness to the Grail,
a symbolical source of physical and spiritual life.   The
chalice rests upon the mercurial symbol.   The adjacent
double circle is again expressive of the dual nature of the
influences at work in producing metals and the Philosopher's Stone, the last of which is denoted by the orb.
The two-headed eagle and the lion represent Wind and
Earth, and the interlinked star with seven points symbolises the seven metals of earth corresponding to the
seven planets in the firmament.   The pointing hand,
with sleeved forearm, sometimes occurs on moon talis-
mans.[84]   The encircling legend runs thus: ' Visit the
inward parts of earth; by rectifying thou shalt find the
hidden Stone '.   The initial letters of the Latin words
form an acrostic, which indicates that the searcher must
use ' vitriol ' in his quest: not any common vitriol, of
course, but ' vitriol of the Sages '.   An accompanying
Latin epigram in *Viridarium Chymicum* (p. 270) states
that ' the wandering planets are beheld in the sky; earth
is equal to them with her yield of metals.   The Sun is
father to the Stone, wandering Cynthia [the moon] its
mother, Wind bore the child in its womb, Earth gave it
food.'   This is a direct allusion to the precepts of Hermes
(p. 54), and the design is often associated in alchemical
works with the Emerald Table.

§ 14. The Religious Element

Since alchemy is said to have arisen in the temples of
ancient Egypt and China, it is not surprising to find that
alchemical ideas have been closely bound up throughout
the ages with religious beliefs.   In the early centuries of
alchemy such ideas were derived largely from forms of
religion practised by Jews, Christians, Gnostics, and Neo-
platonists; but there are alchemical strands which may
be traced back to much earlier and more primitive forms

of religion than those of the Alexandrian era. To take an outstanding example, a slight acquaintance with alchemical writings and emblems is sufficient to indicate the importance which was attached in alchemy to the serpent. This leading tendency finds a notable expression in the Figures of Abraham the Jew (Plate 7); and these representations are symptomatic, for throughout alchemical symbolism the serpent, snake, or dragon, occurs with remarkable frequency.

The most familiar form of the talismanic serpent is the caduceus (Plates 18 (i) and 30); this was usually, but not always, associated with Hermes, or Mercury (p. 5). The normal form is a winged wand entwined by two twigs or serpents, but sometimes the wings are absent. It is said that the original form, attributed to Thoth, was a cross, symbolising the four elements proceeding from a common centre. The central stem of the alchemical caduceus was sometimes held to consist of 'gold of the Philosophers' (p. 102), and the two serpents were said to represent either the male and female or the fixed and volatile principles.

There is little doubt that alchemy inherited the symbol of the serpent from the early mythological systems disposed around the cult of serpent-worship. This cult, in turn, was closely bound up with sun-worship and phallism. Serpent-worship, which was very widespread in the ancient world, possibly arose in Babylonia, where the worship of the sun and the serpent appears to have been the earliest form of religion. Ea, or Hea, god of wisdom and father of the Babylonian sun-god, bore the emblem of a seven-headed serpent. Originally held sacred as the symbol of the sun-gods, the serpent eventually achieved independent divine rank. In the ancient world, ' it entered into the mythology of every nation, consecrated almost every temple, symbolised almost every deity, was imagined in the heavens [p. 100], stamped on the earth, and ruled in the realms of everlasting sorrow '. There is a theory that Stonehenge was a solar temple, which, according to Sir Norman Lockyer's astronomical calculations, may have been erected about 1680 B.C.; and it has even been conjectured that the still older stone circles of Avebury, on the Marlborough downs, formed part of a great serpentine temple and design, with the head resting on Hakpen Hill (*hak*, hook or serpent; *pen*, hill or head).

The serpent, or dragon, in mythology as also in alchemy, was susceptible of numerous interpretations. ' More subtil than any beast of the field ',[85] it was symbolical of divine wisdom; * also of power and creative energy; of time and eternity; of life, immortality, and regeneration. Moreover, the serpent was used as a solar emblem, a phallic emblem, and an emblem of the earth; sometimes also it symbolised the hermaphroditic principle (p. 102). In some cosmogonies, with the egg, the symbol of fertility, we find the serpent, the creature of fertile chthonian gods. Further, it is of interest that the Sun-god and Moon-goddess, or Great Father and Great Mother (p. 20), were figured not only as lion and lioness, bull and cow, etc., but also as male and female serpent. In alchemy, the dragon, or serpent, is often used as a generative or sexual symbol. Sometimes male and female serpents or dragons are pictured as devouring or destroying each other, thereby giving rise to a glorified dragon, typifying the Philosopher's Stone, or transmutation; the same symbol may also denote putrefaction (p. 138).

Winged and wingless serpents or dragons symbolise the volatile and fixed principles (mercury and sulphur) respectively; three serpents, the three principles (mercury, sulphur, salt); and a serpent nailed to a cross, the fixation of the volatile. Flamel likened the great stench of dragons to that of sophic mercury and sulphur.

The widespread veneration paid to the serpent appears to have been due largely to its mysterious nature and its unusual form and habits. Thus, the annual shedding of its skin, and the apparent accompaniment of rejuvenation,

---

* In Greek art, Athena often carries the Gorgon's snaky head on her aegis.

were linked with the conception of immortality and with that idea of regeneration or new birth which was a characteristic of the phallic system. The same phenomenon was associated with the supposed endless succession of forms in Nature, and with the annual dissolution and renovation of the seasons, due to the sun. Among other supposed peculiarities of the serpent were great weight and slipperiness, reminiscent of the elements earth and water, respectively, and its reputed power of achieving the eternal cycle by devouring its own body.

In ancient Egypt, a serpent biting its tail was a symbol of eternity, or of the universe. This device, known as the Ouroboros* Serpent, occurs on ancient Egyptian papyri dating back as far as the sixteenth century B.C. It is found also in Greek manuscripts of the Alexandrian period: for example, the tail-piece on page 117 is reproduced from the *Chrysopoeia* ('Gold-making') of Cleopatra. The enclosed words, ἓν τὸ πᾶν (' the all is one '), refer to the Platonic idea of the unity of matter, which formed one of the fundamentals of alchemical theory. The circle formed by the serpent embodies the same general idea, and possibly also the more particular alchemical conception of circulation within the closed vessel (p. 15). ' The serpent is one who has the poison '—that is, possibly, the Philosopher's Stone—runs a Greek inscription on a drawing associated with Cleopatra's representation of the Ouroboros Serpent; and a further inscription states, ' One is all, and by it all, and to it all, and if one does not contain all, all is nought '.[86]

In the Alexandrian era, ' Greeks, Jews, Egyptians and Christians ', to quote the words of Hopkins, [87] ' differing from one another in many ways . . . all believed in a god . . . yet also were they all troubled by one common mystery. . . . How could God be far away and all-powerful and yet near to man and sympathetic? Historically, the first solution to this question was offered by the Alexandrian Jews in imagining a mediator as agent of

* ' Tail-eater.'

108

God and yet close to man. . . . Their imagined inter-
mediary they called " Sophia " or Wisdom, who bridges
the chasm between God and man. One well-known
result of this idealisation of Sophia or Logos was the
Wisdom of Solomon, about 100 B.C., a book now forming
a portion of our Apocrypha.'

Davis,[88] writing in a similar vein, remarks that ' from
earliest times the mind of man has grasped tangible and
formulated intangible Nature by means of two opposite
qualities and a third by which the opposites are mediated,
reconciled and included. The notion of the Trinity,
Father, Son and Godhead, representing active, passive
and spirit, respectively, antedates history. The Egyptian
'ankh, the symbol of life, is a combination of male and
female. By slight modification it became the cross, still
the symbol of life, upon which, before Christianity, the
irony of the Jews saw fit to nail their criminals until they
were dead. Nor has the human mind yet finished with
this sort of thinking. The metaphysics of Hegel and
Royce are evidences of it, as is the Langmuir atom from
which electrons take their being and in which positive and
negative are neutralised.'

' One-two-three: positive-negative-neutral. This is
the frame upon which must be stretched whatever it is we
wish to understand,' states the same writer [89] in a contribution on primitive science, ' and this is the frame upon
which from time immemorial the mind of man has stretched
it. It is the mode of all knowing. If we wish to know a
man, we must consider him as active subject (What does
he think of himself?), as passive object (What do his
fellows think of him?), and, thirdly, the man as he really is.
The philosophy of Hegel is but a reiterated insistence
upon the three—the Thing, its Own Other, and the Thing-
in-Itself. If we wish to think of God, we are obliged to
think of Him as Active Power, as passive object, that is, as
that which possesses passivity and undergoes the Passion,
and thirdly as the Spirit which determines and terminates
the action, the Soul, the Body, and the Spirit.'

PRELUDE TO CHEMISTRY

Alchemical literature, much of which was written by men of deep religious convictions, contains innumerable references to religious beliefs and doctrines, of which a few typical examples may now be adduced. A definite parallel between the Christian mystery of the Godhead and the alchemical mystery of the Stone is expressed in the following rendering [90] of the prayer of ' the learned and rare Philosopher of our Nation ', George Ripley (1415–1490), canon of Bridlington, in the prologue to his popular work, *The Compound of Alchymie . . . Conteining twelve Gates*: ' O Unity in the substance, and Trinity in the Godhead. . . . As thou didst make all things out of *one* chaos, so let me be skilled to evolve our microcosm out of *one* substance in its three aspects of Magnesia, Sulphur, and Mercury.' Ripley thus pictures the Stone as a triune microcosm (*cf.* p. 133), and likens the alchemical mystery to the Christian mystery of the Godhead.

The same conception is found in an old English poem happily preserved by Ashmole,[91] in which the Stone is endowed with a heavenly origin. The poem relates, in the quaint language of an earlier day, how the blessed Stone ' Fro Heven wase sende downe to *Solomon* '. Its three constituents are likened to the gifts of gold, and frankincense, and myrrh, which the three Magi brought to Bethlehem:

Aurum betokeneth heer, owre Bodi than,
The wych was brought to God and Man.

And Tus alleso owre Soul of lyfe,
Wyth Myrham owre Mercurye that ys hys Wyfe.

Here be the thre namys fayre and good
And alle thaye ben but one in mode.

Lyke as the Trinite ys but on,
Ryght so conclude the *Phylosofeers Stone*.

An accompanying illustration (Fig. 10) depicts Sol and Luna enclosed in a Hermetic Vase; under the animating influence of a breath from heaven (p. 85), the Philosopher's Stone is formed from these ingredients, and seven metallic streams flow from the Vase to the earth below, symbolised by the toad. The design includes

Fig. 10.—Heavenly origin of the Stone.
From *Theatrum Chemicum Britannicum*, Ashmole, 1652.

also the dragons of sophic sulphur and mercury, figures of the ' *Red Man* and hys *Whyte Wyfe* ', and a reference to spirit, soul, and body.

Basil Valentine [92] makes a similar comparison in his ' Allegorical expressions betwixt the Holy Trinity and the

Philosophers stone '; and the Conclusion of *Bloomefields Blossoms* [93] contains the injunction,

First I say in the name of the holy Trinity,
Looke that thou joyne in One, Persons Three.

A further curious parallel is drawn by Basil Valentine: [94] ' As the four Evangelists are witnesses of the New Testament and Covenant; so they are a type and sure testimony of the four elements, that the Earth is created after the holy Heaven '.

A heavenly influence was often held to dominate the production of the Philosopher's Stone (p. 143); this is evident, for example, from Fig. 10. An elaborate genealogical tree of the Philosopher's Stone, dating from the late fifteenth century, derives the Stone indirectly from God, through the intervention of Nature, according to the Neo-Platonic concept.[95] The writer of *Gloria Mundi* (' The Glory of the World ') likens the Stone to the biblical ' stone which the builders rejected ', in the following remarkable passage: [96] ' The Stone is cast away and rejected by all. Indeed it is the Stone which the builders of Solomon disallowed. But if it be prepared in the right way, it is a pearl without price, and, indeed, the earthly antitype of Christ, the heavenly Corner Stone. As Christ was despised and rejected in this world by the people of the Jews, and nevertheless was more precious than heaven and earth; so it is with our Stone among earthly things.'

Gloria Mundi refers the ultimate origin of alchemy to God; for the title-page runs: ' The Glory of the World; or, Table of Paradise; that is to say, A True Account of the Ancient Science which Adam learned from God Himself; which Noah, Abraham, and Solomon held as one of the Greatest Gifts of God '. The text of the work [97] contains the supplementary statement: ' When Adam had learned the mystery out of God's own mouth, he kept it a strict secret from all his sons, until at length, towards the close of his life, he obtained leave from God to make the preparation of the Stone known to his son Seth '.

The close relationships which existed between alchemy and religion in mediaeval times and later may be illustrated in another way. Elias Ashmole, writing in 1652, recorded the existence at that period of an alchemical painting upon an arched wall in Westminster Abbey (p. 173). In the same place [98] he gives a detailed description of a former window in St. Margaret's Church, Westminster, which symbolised in coloured designs the whole process of preparing the Philosopher's Stone. The designs included man and woman, sun and moon, ' a faire large red rose ', a bright yellow glory, and a man clad in red, holding a white stone in his left hand and a red stone in his right hand. It appears from this evidence that Westminster Abbey, like Notre-Dame, may have been at one time frequented by alchemists—an association which again draws attention to the connection between the Christian mystery and the alchemical mystery. Ripley, indeed, talks of ' *Westminster* Church, To whych these *Phylosophers* do haunte '.[99]

It is pertinent to the subject to remark at this point that alchemical ideas found expression in stone, as well as in paintings and coloured glass. De Givry,[100] for example, has described the alchemical significance of the original statue of St. Marcellus, ' the principal alchemic hieroglyph of Notre-Dame ', to be seen on the central pillar of the southern doorway of the main front. The original fourteenth-century statue is preserved in the Musée de Cluny. In this design a dragon (sophic mercury) is shown climbing out of the athanor (furnace), with the Bishop's crozier thrust into its throat—signifying ' that a shaft of celestial light is necessary for kindling the fire of the athanor ', in the preparation of the Philosopher's Stone.

Lacroix,[101] also, comments upon ' those wonderful Gothic monuments in which the statuary has represented a mass of figures, sacred and profane, real and imaginary, and which give one the impression of being a book of alchemy, written with a chisel upon stone '.

Alchemical symbols and emblems were often endowed with a twofold significance, as exemplified by the Figures of Abraham (p. 62), which, to use words ascribed to Flamel himself,[102] 'may represent *two things*, according to the capacity and understanding of them that behold them: First, the *mysteries* of our future and undoubted *Resurrection*, at the day of Iudgement, and comming of good *Iesus*, (whom may it please to have mercy upon us) a Historie which is well agreeing to a *Churchyard*. And secondly, they may signifie to them, which are skilled in Naturall *Philosophy*, all the principall and necessary operations of the *Maistery*. These *Hieroglyphicke figures* shall serve as two wayes to leade into the heavenly life: the first and most open sence, teaching the sacred *Mysteries* of our salvation; (as I will shew heereafter) the other teaching every man, that hath any small understanding in the *Stone*, the lineary way of the worke; which being perfected by any one, the change of evill into good, takes away from him the roote of all sinne (which is *covetousnesse*) making him liberall, gentle, pious, religious, and fearing God, how evill soever hee was before, for from thence forward, hee is continually ravished, with the great grace and mercy which hee hath obtained from God, and with the profoundnesse of his Divine & admirable works.'

Sometimes this dual interpretation of alchemical emblems was carried to great lengths. For example, the eleventh book of Michael Maier's *Symbola Aureae Mensaē* of 1617, described below (p. 222), deals with a certain Melchior Cibinensis,* an alleged Hungarian 'hero' of alchemy. The accompanying emblem (Plate 19) refers ostensibly to the alchemical operation of cibation (p. 139): the infant deriving sustenance from its nurse, the earth, is an image of the Stone—whose mother, in the words of the fourth precept of Hermes (p. 54), is the moon. Stolcius (p. 258) comments upon this emblem in the following terms: 'Melchior is famed as born in the borders of

---

* Cibinensis, *i.e.* of Szeben or Sibiu (Hermannstadt), in Transylvania.

Hungary; although a priest, he won golden gifts. Under the form of the Mass he described the Stone, whether well, judge each for himself. As a tender babe is at first fed with snow-white milk, so this Stone must be nurtured with pure milk.' In discussing this emblem, Waite [103] suggests that Melchior was 'invented' by Maier as a peg upon which to hang the cloak of certain Christian mysteries. Maier describes Melchior as a Christian priest who has graduated in 'the hidden mysteries of the hidden science'. In the emblem he wears elaborate sacerdotal vestments, and is engaged in celebrating Mass. 'In the person of this exponent of the debate,' writes Waite, 'Maier represents the Mass as a work of the hidden science and the sanctuary of its Mysteries, which are those of the Philosopher's Stone. It is said also (1) that in the Sacrament of the Altar are concealed the most profound secrets of spiritual Alchemy; (2) that the perfection of the Great Work is the birth of the Philosopher's Stone in the Sacred Nativity; (3) that its sublimation is the Divine Life and Passion; (4) that the black state represents the death on Calvary; and (5) that the perfection of the red state corresponds to the resurrection of Easter and the Divine Life thereafter.' Maier adds that 'these earthly things are a picture of those which are heavenly'.

Thus, even the chemical changes occurring in alchemical processes were liable to a twofold interpretation. A still more definite example of the kind is to be found in Redgrove's [104] correlation of the three main colour stages of the Great Work (p. 135) with the threefold division of the regenerative process of religious mysticism: the black stage is held to correspond to 'the dark night of the soul', the white to 'the morning light of a new intelligence', and the red to 'the contemplative life of love'. According to this view, the alchemists 'derived their beliefs concerning the colours, and other peculiarities of each stage in the supposed chemical process [of preparing the Philosopher's Stone], from the characteristics of each stage in the psychological process according to mystical theology '.

§ 15. Conclusion

All the works which have found mention in this brief survey of the literature of alchemy, and many others, have a definite interest for the modern reader and student. It must be admitted, however, that a large proportion of alchemical literature is dull, arid, or unintelligible; some of the mystical and visionary writings, indeed, bear the marks of incipient lunacy.

A general impression of alchemical literature may be gained by an examination of such representative compilations as Zetzner's *Theatrum Chemicum* (' Chemical Theatre '), published in six volumes at Strassburg in 1659, and Manget's *Bibliotheca Chemica Curiosa* (' The Careful Enquirer's Chemical Library '), printed in two large folio volumes at Geneva in 1702; these two ' omnibus ' works convey also some idea of the enormous output of alchemical writings, for between them they contain a selection of more than three hundred tracts occupying nearly seven thousand pages. Alchemy was much given to the issue of collections of tracts, and many of the writings are not to be found as independent publications. Of the many collections, one of the most readable is the *Musaeum Hermeticum* (' Hermetic Museum '), and this finds further mention below (p. 166). According to Waite,[105] the *Musaeum Hermeticum* was designed ' to supply in a compact form a representative collection of the more brief and less ancient alchemical writers; in this respect, it may be regarded as a supplement to those large storehouses of Hermetic learning such as the *Theatrum Chemicum*, and that scarcely less colossal of Mangetus, the *Bibliotheca Chemica Curiosa*, which are largely concerned with the cream of the archaic literature, with the works of Geber and the adepts of the school of Arabia, with the writings attributed to Hermes, with those of Raymond

Lully, Arnold de Villa Nova, Bernard Trevisan, and others '.

A still more comprehensive view of alchemical literature is afforded by Ferguson's classical *Bibliotheca Chemica*.[18]   This monumental work, consisting of 1085 quarto pages, is a catalogue, with brief annotations and biographical notes, of the James Young collection of some 1300 pamphlets and volumes, dealing mainly with alchemy.[106]   A familiarity with the plan and scope of this annotated list of a single collection of alchemical writings suffices to create a striking impression of the almost illimitable extent of a field of literature which embraces a period of some fifteen hundred years.

The Ouroboros Serpent.
From the *Chrysopoeia* of
Cleopatra.

CHAPTER III

THE PHILOSOPHER'S STONE

*Every theory which urges men to labour and research, which excites
acuteness and sustains perseverance, is a gain to science; for it is
labour and research which lead to discoveries.*　　　LIEBIG

§ 1. ORIGIN AND NATURE OF THE QUEST

THE history of science contains no parallel
to the quest of the Philosopher's Stone;
it contains nothing else so impressive or
romantic. The achievement of the Stone
was the great and final goal of alchemy.
In its fullest aspect, the quest was dominated by a singularly noble ideal, for it
was imperfect man's search after perfection. To most of
its seekers, however, the Grand Magisterium and Elixir
was merely a magic source of wealth, health, and long life.
'Gold', says Goethe, 'gives power; without health there
is no enjoyment, and longevity here takes the place of
immortality.' With such considerations in mind, and inspired by their vivid faith in the miraculous virtues of the
imaginary Stone, generation after generation of alchemists,
during an epoch of more than a thousand years, devoted
their lives and treasures to the pursuit of the greatest *ignis
fatuus* which the world has ever known. These ardent
'labourers in the fire' pressed forward in an unceasing
stream, year after year, decade after decade, century after
century, to maintain an unwearying assault upon the Alchemical Citadel; but their struggles met with no reward

during the long age of alchemy. Nevertheless, from these tedious labours, seemingly so barren, sprang the seed which bourgeoned in a later age into the science of chemistry. 'The most lively imagination', wrote Liebig,[1] ' is not capable of devising a thought which could have acted more powerfully and constantly on the minds and faculties of men, than that very idea of the Philosopher's Stone. Without this idea, chemistry would not now stand in its present perfection. . . . In order to know that the Philosopher's Stone did not really exist, it was indispensable that every substance accessible . . . should be observed and examined. . . . But it is precisely in this that we perceive the almost miraculous influence of the idea. The strength of opinion could not be broken till science had reached a certain stage of development.'

It is not known where or when the idea of the Philosopher's Stone originated. The evidence — so far as evidence can be said to exist — seems to point to Alexandria, in the early centuries of the Christian era (p. 12); but it is remarkable that the belief in a transmuting process was common to the Occidental and Oriental civilisations (p. 6). The Occidental idea of the Stone is closely allied with the profound belief in magic which was cherished in ancient Egypt, and later by the Neo-Platonists and the Arabs; it is also in keeping with the early theories of the constitution of matter and with the mystical beliefs concerning the regeneration of man. It is doubtless the Stone which finds a place in the Emerald Table of Hermes under the designation of 'the father of perfection throughout the world '.

The fundamental idea of the Philosopher's Stone was expressed by Arnold of Villanova[2] in the following terms: ' That there abides in nature a certain pure matter, which, being discovered and brought by art to perfection, converts to itself proportionally all imperfect bodies that it touches '. This idea, in one form or another, was undoubtedly the main source of inspiration of alchemists of all kinds, from the most exalted adept to the meanest puffer. Thus, Pernety,[3] as a result of a close study of alchemical writings, defined alchemy as ' a science, and the art of making a fermentative powder, which transmutes imperfect metals into gold, and which serves as a universal remedy for all the ills of men, animals, and plants'. It is surprising that this definition was published so late as 1787.

The supposed possibility of transmutation was bound up with the alchemical doctrine of the unity of all things; and this doctrine, in turn, was a heritage from the Ionian philosophers, who held that all bodies possess a common basis in the *prima materia*, or single primordial matter: even the four elements of Aristotle were held to be interconvertible (p. 10). Moreover, the sulphur-mercury theory, descended from the ancient doctrine of the Two Contraries (p. 19), provided a common basis for the metals. The possibility of transmutation was thus a logical consequence of alchemical theory. Quite apart from theoretical considerations, however, the alchemists regarded transmutation as a self-evident truth. Why, they asked, should Nature find any difficulty in making gold and silver from lead or mercury, when she is able to carry out the much more unlikely transformations of grass and rain-water into flowers, and of wheat and hay into the bones and muscles of animals?

Nor were the alchemists at a loss for practical evidence in support of their views upon the transmutation of metals. In their eyes, such processes could be proved experimentally in many ways, as well as being justified on theoretical grounds. For example, the mineral galena (lead sulphide) possesses the colour and lustre of lead, but it is neither malleable nor fusible like lead. When it is heated, however, it disengages sulphureous fumes, acquires the missing properties, and is transformed into lead. Could not lead by further heating, or other treatment, be made to lose more sulphur and be endowed thereby with other properties which would ennoble it

PLATE 27

Consummation of the Great Work.

From *Mutus Liber* in *Bibliotheca Chemica Curiosa*, Manget, 1702. (See p. 159.)

PLATE 28

The washing of Latona with Azoth.

From *Secretioris Naturae Secretorum Scrutinium Chymicum*, Maier, 1687 (reprinted from *Atalanta Fugiens*, Maier, 1618). (See p. 162.)

PLATE 29

'Le Grimoire d'Hypocrate': an Alchemical Interior.

D. Teniers *pinxit*, F. Basan *sculpsit* (Paris).   (See pp. 164, 305.)

PLATE 30

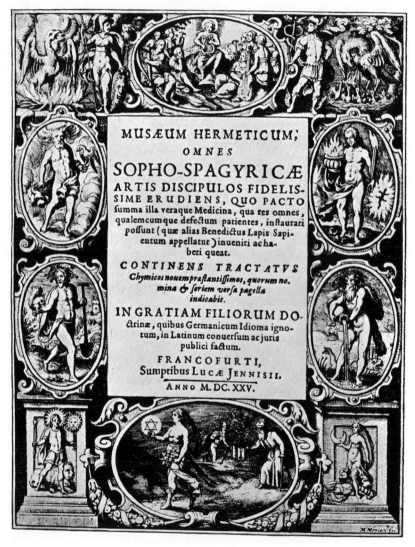

Title-page of *Musaeum Hermeticum*, 1625.

further to silver, and eventually to gold? As it happens, lead from galena often contains considerable quantities of silver, which may be separated from it by further heating in the process of cupellation. Such observations, suggesting an actual conversion of part of the lead into silver, confirmed the faith of the transmutationists. Moreover, there is no doubt that experimental evidence of this kind played an important part in the development—from Aristotle's theory of smoky and vaporous exhalations—of the celebrated sulphur-mercury theory of the origin of metals (p. 18). Mercury and sulphur typified metallic character and inflammability respectively: only after conversion into its basic or primal ' mercury ', or ' mercurial first matter ', could a metal be made free to pass into another form.

Undoubtedly there is a vein of reason running through the mountainous accumulation of literature which deals with transmutation and the Philosopher's Stone; but the vein is usually so deeply embedded as to be undiscernible without the aid of a mental divining rod.

§ 2. THE ELIXIR OF LIFE

In due course, as a further consequence of their fundamental belief in the unity of all things, the alchemists came to regard the medicine of the metals as the medicine of man also. The Philosopher's Stone, under such names as the Elixir of Life, or Grand Elixir, was depicted as a panacea for all human ills, capable also of restoring youthfulness and prolonging life.

Liebig [4] assigned this extension of the powers of the imaginary Stone to the thirteenth century; but more recent researches (p. 6) show that the idea was familiar in China—where it probably originated—and presumably also in the West, at a much earlier date. Indeed, the concept of a potent medicine possessing miraculous powers is one of the earliest and most important features of Chinese alchemy. Davis [5] emphasises the important fact that Chinese alchemy incorporated a notion which was unknown in European alchemy, namely, the assumed existence of benevolent supernatural beings (*hsien*) who had achieved immortality by chemical means. Further, the *Ts'an T'ing Ch'i* of Wei Po-yang, dating from the second century A.D. (p. 7), is really a treatise on the preparation of the ' pill of immortality '. Wei Po-yang states that 'men of art' by feeding on gold attain longevity. Pills made of the *huan-tan*, or Returned Medicine, he says, ' are extremely efficacious, although their individual size is so small that they occupy only the point of a knife or the edge of a spatula '.

Force is lent to this statement by a story of Wei Po-yang which is related[6] in the *Lieh Hsien Ch'üan chuan*, or ' Complete Biographies of the Immortals '. The Master resorted to the mountains with three disciples, whose faith he wished to test, and his white dog. Having prepared a ' gold medicine ', he fed it to the dog. The animal forthwith expired. Wei Po-yang observed blandly, ' Doesn't this show that the divine light has not been attained? If we take it ourselves, I am afraid we shall go the same way as the dog. What is to be done? ' The only suggestion of the disciples was, ' Would you take it yourself, Sir? ' The Master replied, ' I have abandoned the worldly route and forsaken my home to come here. I should be ashamed to return if I could not attain the *hsien* (immortal). So, to live without taking the medicine would be just the same as to die of the medicine. I must take it.' He did so—and immediately, as he had surmised, went the same way as the dog. His faithful disciple, Yü, followed suit. The remaining two disciples, however, ' bethought themselves and went ' to get burial supplies for their departed colleagues. Wei Po-yang then revived, since the medicine contained only sufficient impurity to cause temporary death. ' He placed some of the well-concocted medicine in the mouth of the disciple and in the mouth of the dog. In a few moments they both revived. He took the disciple, whose name was Yü, and the dog, and went the way of the immortals. By a wood-cutter whom they met, he sent a letter of thanks to the two disciples. The two disciples were filled with regret when they read the letter.'

A quaint drawing, reproduced by Wu and Davis in *Isis*,[7] from a Japanese woodcut about 500 years old, depicts Wei Po-yang, ' the father of Chinese alchemy ', with his white dog and the faithful disciple, Yü; the furnace, or *ting*, in which the medicine (*huan-tan*) was prepared, is also shown (Plate 20). Another similar drawing illustrates the picturesque statement that the fowls and barnyard animals which ate of the residues from such vessels ascended into the clouds.

The later Western alchemists made similar and almost equally picturesque claims concerning the potency of the Elixir of Life. ' Our Medicine ', wrote Arnold of Villanova,[8] ' has also power to heal all infirmity and diseases, both of inflammation and debility; it turns an old man into a youth. If the illness be of one month's standing, it may be cured in a day; if of one year's standing, it may be healed in twelve days; if of many years' standing, it may be healed in a month. Hence this Medicine is not without reason prized above all other treasures that this world affords.' Ripley [9] made a similar claim in the epistle to his *Compound of Alchymie:* ' *Then will that Medicine heale all manner Infirmitie, And turne all Mettalls to* Sonne & Moone *most perfectly* '.

Isaac of Holland, writing apparently in the sixteenth century, enlarges upon the medicinal virtues of the Stone in his *Opus Saturni*, and prescribes a dose of the size of a grain of wheat every nine days: ' the patient shall think he is no longer a man, but a spirit ', says Isaac. ' He shall feel as if he were nine days in Paradise, and living on its fruits.' Such effects were perhaps partly due to the spirit of wine, in which the Stone was said to be dissolved in preparing the Elixir of Life. Salomon Trismosin,[10] writing in the same vein towards the end of the sixteenth century, describes his rejuvenation by means of half a grain of the Stone; moreover, he asserts cheerfully that in a similar way, with the aid of a medicine from the ' Red Lion ', he had succeeded in restoring to perfect youthfulness ladies whose venerable forms had witnessed the passage of between seventy and ninety summers!   It would have been easy, he adds, for him to prolong his own life until the last day, by using this potent medicine; but apparently he had no desire to do so.   Figuier[11] mentions, however, that according to numerous enthusiastic adepts, Nicolas Flamel and Perrenelle his wife achieved perennial youth and migrated to Asia; here, they were said to be living as late as the eighteenth century, more than three hundred years after their presumed deaths!

Even the longevity of the patriarchs was attributed to their use of the Stone.[12]   ' Unless Adam had possessed the knowledge of this great mystery,' remarks the writer of *Gloria Mundi*,[13] ' he would not have been able to prolong his life to the age of three hundred (let alone nine hundred) years.'

Norton and others held also that the Stone perfected man's mental faculties and moral nature.   ' I assure you ', exclaims Mammon in *The Alchemist*,[14]

> He that has once the flower of the sun,
> The perfect ruby, which he calls elixir,
> Not only can do that, but, by its virtue
> Can confer honour, love, respect, long life,
> Give safety, valour, yea, and victory
> To whom he will.   In eight and twenty days
> I'll make an old man, of fourscore, a child.

Surly. No doubt—he's that already!

Mammon.   Nay, I mean,
> Restore his years, renew him, like an eagle,
> To the fifth age; make him get sons and daughters,
> Young giants; as our philosophers have done
> (The ancient patriarchs afore the flood)
> But taking, once a week, on a knife's point,
> The quantity of a grain of mustard of it . . .
> I'll undertake, withal to fright the plague
> Out o' the kingdom in three months.

THE PHILOSOPHER'S STONE

§ 3. OTHER VIRTUES OF THE STONE

According to Trismosin,[15] the Stone possessed four principal virtues; for besides bringing perfection to the metals and healing to man, it could change all base stones into precious ones, and soften every kind of glass.   Some writers held that the perfect Stone or Elixir was composed of a quintessence of three Stones, namely, the mineral, vegetable, and animal.   According to Arnold of Villanova, the Stone was ' mineral because it is compounded of mineral things, vegetable because it lives and vegetates, and animal because it has a body, a soul, and a spirit, like animals '.   Elias Ashmole, in the introduction to his *Theatrum Chemicum Britannicum*,[16] took a still wider view of the subject, in particular of the ' stupendious *and* Immense *things* ' which could ' *bee performed by vertue of the* Philosophers Mercury, *of which a* Taste *onely and no more* ':

The Mineral Stone, he says, has the power of transmuting any imperfect earthy matter into its utmost degree of perfection; that is, to convert the basest of metals into perfect gold and silver, flints into all manner of precious stones, and so forth.   The Vegetable Stone controls and stimulates the growth of trees, plants, and flowers: therefore it is not unlikely that the Glastonbury Thorn, ' *so greatly fam'd for shooting forth* leaves *and* Flowers *at* Christmas ', may be an experiment ' *made of the* Vegitable Stone '.   Moreover, ' *if the* Lower *part be expos'd abroad in a* dark Night, Birds *will repaire to (and circulate about) it, as a* Fly *round a* Candle, *and submit themselves to the* Captivity *of the* Hand '.

There are even greater wonders to follow, for the Magical or Prospective Stone makes it ' *possible to discover any* Person *in what part of the* World *soever, although never so secretly concealed or hid, in* Chambers, Closets, *or* Cavernes *of the* Earth. . . . *Nay more, It enables Man to* understand *the language of the* Creatures, *as the* Chirping *of* Birds, Lowing *of* Beasts, *&c.*'   More remarkable still, this

Stone ' *is not any wayes* Necromanticall, *or* Devilish; *but* easy, wonderous easy, Naturall *and* Honest '.

There is finally the Angelical Stone, which is so subtle that it ' *can neither be* seene, felt, *or* weighed; *but* Tasted *only* '. It enables one to live a long time without food, and gives its possessor the power of conversing with angels by dreams and revelations. Ashmole adds that ' *after* Hermes *had once obtained the* Knowledge *of this* Stone, *he gave over the use of all other* Stones, *and therein only delighted*: Moses, *and* Solomon, (*together with* Hermes *were the only three, that*) *excelled in the* Knowledge *thereof, and who therewith wrought* Wonders '.

The fact that such fantastic statements as these could be accepted and solemnly repeated by a cultured contemporary of Robert Boyle throws into strong relief the amazing credulity which prevailed, even among intelligent and educated men, so late as the middle of the seventeenth century.

§ 4. THE ALKAHEST

Bound up to some extent with the quest of the Philosopher's Stone, was the search for the so-called Alkahest, an imagined universal solvent, which, according to Paracelsus and van Helmont, had the power of converting all bodies into their liquid primary matter.[17] This idea appears to derive from Paracelsus, who coined the word *alkahest* from the German *allgeist*, or ' all-spirit '.[18] Liquefaction by means of this magic solvent was held to differ from ordinary processes of solution, by reason of its gentle and non-corrosive nature. Acids and other corrosive solvents were likened in their effects to ordinary or ' elementary ' fire, which was said to destroy and kill, instead of vivifying, like the Fire of the Sages (p. 143). The possession of the Alkahest was thus held, in some of the later alchemical writings, to constitute another fundamental difference between the true Hermetic philosophers and the puffers, or chymists. ' The chymists destroy ', stated, in effect, the esoteric fraternity, 'we build; they burn with fire, we with water; they kill, we resuscitate; they wash with water, we with fire'. The existence of the Alkahest was accepted by such outstanding practical workers as van Helmont and Glauber; van Helmont described an experiment made with a specimen of the potent fluid which he obtained from a mysterious adept, and Glauber claimed that his *sal mirabile* (sodium sulphate) was the long-awaited material. It is remarkable that his use of a glass container for the universal solvent failed to strike van Helmont as a paradoxical proceeding. The strange belief in the existence and powers of this imaginary solvent persisted into the eighteenth century, and eventually the Alkahest left the alchemical stage not long before the Philosopher's Stone.

§ 5. Designations and Descriptions of the Stone

The designations of the Stone and the accounts of its appearance are innumerable: many alchemical writers, indeed, appear to have regarded it both as a duty and a mark of their originality to advance a new name for the Stone, or to provide it with a new antithetical description. Although sometimes described as a liquid, it is more often mentioned as a powder, which, according to its white or red colour, is able to transmute base metals into silver or gold. The powder was said to be fixed, heavy, and of a pleasant odour; sometimes it was called the powder of projection. The Stone was often assumed to exist in different orders, or degrees of perfection.

In course of time the alchemists assembled an enormous variety of names for the imaginary Stone. Unfortunately they enhanced the general confusion of alchemical literature by using many of these terms for other purposes, in particular to denote sophic mercury, sophic sulphur (p. 103), or other so-called materials of the Great Work, or Preparation of the Stone.

There is, for instance, a rare little book [19] entitled

The names of the Philosophers Stone, ' collected by William Gratacolle ' and translated into English ' by the Paines and Care of H. P.' (London, 1652), which contains a list of more than 170 synonyms. Among these are ' brasse of Philosophers, virgins milke, a high man with a Sallet [helmet], the shaddow of the Sun, a crowne overcomming a cloud, the bark of the Sea, the water of Sulphur, the spittle of Lune, venome, Azot, whitenesse, blacker than black, dry water, the lesse world, and almost with other infinite names of pleasure '.

Pernety [20] gave a partial alphabetical list containing about 600 names which were used either for the Stone or for the related materials: these range from Absemir, Adam, and Adrop to Zotichon, Zumech, and Zumelazuli. In *Bloomefields Blossoms*,[21] ' the blessed Stone ' is called ' Our greate *Elixer* most high of price, Our *Azot*, our *Basaliske*, our *Adrop*, and our *Cocatrice* . . . our *Antimony* and our *Red Lead* . . . our Crowne of Glory and Diadem of our head . . . our *Lyon* . . . the *Salamander* by the fire living . . . the *Metalline Menstruall*, it is ever the same . . .

> ' Some call it also a *Substance exuberate*,
> Some call it *Mercury* of Metalline essence,
> Some *Limus deserti* from his body evacuate,
> Some the *Eagle flying* from the North with violence:
> Some call it a *Toade* for his great vehemence.'

Among other names, chosen at random, are Phoenix, Serpent's Brother, Universal Medicine, Yolk of the Egg, Carbunculus, Jud he voph hé, Grand Magistery or Magisterium, Xit, Stone of the Sages, Serpent eating his Tail, Universal Essence, Bath or Fountain of the King, and Blood of the Salamander. Many of the terms, such as the Red Lion, or Red Tincture, were derived from the supposed colour of the Stone; others, such as Water Stone of the Wise or Sophic Hydrolith, from its supposed fluid or mercurial constituent.

Typical of the many mystical and antithetical descriptions of the Stone is the following, which has been attributed to Zosimos the Panopolitan : ' In speaking of the Philosopher's Stone, receive this stone which is not a stone, a precious thing which has no value, a thing of many shapes which has no shapes, this unknown which is known of all '. Mercer [22] points out that such contradictory descriptions may infer that the Stone ' is at once devoid of all specific qualities, and yet potentially possesses them all. Looked at from this point of view, there is something Hegelian in the rhapsody. Or again . . . the Quintessence is dear and glorious to him who knows it and uses it, vile to him who is ignorant of it; finite and specific for the one, infinite and indeterminate for the other.'

Philalethes states in *A Brief Guide to the Celestial Ruby* [23] that the Philosopher's Stone ' is called a stone, not because it is like a stone, but only because, by virtue of its fixed nature, it resists the action of fire as successfully as any stone. In species it is gold, more pure than the purest. . . . If we say that its nature is spiritual, it would be no more than the truth ; if we describe it as corporeal, the expression would be equally correct.'

Ripley regarded the Stone as a triune microcosm (p. 110). Basil Valentine depicted it as ' composed out of one, two, three, four, and five ' (p. 208), these numbers indicating, respectively, the primordial matter, the two-fold mercurial substance, the three principles (*tria prima*), the four elements, and the quintessence.

These varied descriptions of the elusive Stone inspired Ben Jonson to sum up the position in the following words : [24]

> Your *Lapis philosophicus*? 'Tis a Stone, and not
> A stone; a spirit, a soul, and a body:
> Which if you do dissolve, it is dissolved;
> If you coagulate, it is coagulated;
> If you make it to fly, it flieth.

§ 6. UBIQUITY OF THE STONE

Emphasis was often laid upon the supposed universal occurrence of the Stone, and this widespread idea was sometimes advanced as a reason for the cryptic nature of the directions given for its preparation. According to a statement dated 1526, occurring in *Gloria Mundi*,[25] the Stone ' is familiar to all men, both young and old, is found in the country, in the village, in the town, in all things created by God; yet it is despised by all. Rich and poor handle it every day. It is cast into the street by servant maids. Children play with it. Yet no one prizes it, though, next to the human soul, it is the most beautiful and the most precious thing upon earth, and has power to pull down kings and princes. Nevertheless, it is esteemed the vilest and meanest of earthly things.'

A striking pictorial illustration of the supposed ubiquity of the Stone occurs in Maier's *Atalanta Fugiens* (Plate 21). As an anonymous writer [26] put it, in 1633:

> Through want of *Skill* and *Reasons* light
>   Men stumble at *Noone day*;
> Whilst buisily our *Stone* they seeke,
>   That lyeth in the *way*.
>
> Who thus do seeke they know not what
>   Is't likely they should *finde*?
> Or hitt the *Marke* whereat they *ayme*
>   Better than can the Blinde?

§ 7. PREPARATION OF THE STONE

The practical operations held to be necessary in the preparation of the Philosopher's Stone were usually known as the processes of the Great Work. A less complete series of operations led, it was supposed, to the Simple Magistery, or Little Work: this was regarded as a White Stone which transmuted the base metals only as far as silver. In the Great Work, Grand Magistery, or Work of the Sages, yielding the Red Stone, full perfection was reached in all respects. Nature could not attain this end without Art, nor could Art encompass it without Nature. Norton wrote in his *Ordinall*: [27]

After all this upon a day
I heard my noble *Master* say,
How that manie men patient and wise,
Found our *White Stone* with Exercise;
After that thei were trewlie tought,
With great labour that *Stone* they Caught;
But few (said he) or scarcely one,
In fifteene Kingdoms had our *Red Stone*.

The Great Work and Little Work were often symbolised by the sun-tree and moon-tree, respectively (Plate 16); the *materia prima* of the Stone was represented in various ways, sometimes by a star, sometimes by the interlaced triangles of Solomon's seal—the sign of ' fiery water ' and the symbol of wisdom.[28]

The innumerable accounts of the preparation and use of the Stone are sometimes pseudo-practical, sometimes mystical; and they present an infinite variety of terms, materials, and operations. It follows that in a sustained course of reading, each tract in this illimitable ocean of alchemical literature heightens the general impression of an irremediable confusion of expression, thought, and method. The diagrammatic summary given below (Fig. 11), dealing with the production of the Philosopher's Stone, is therefore to be recognised merely as a reduction to its simplest terms of one of the many schemes of transmutation which are shadowed with varying degrees of unintelligibility, deliberate or involuntary, in alchemical literature.

The first purification gave the so-called ' primitive materials ', or remote matter, of the Stone: in preparing them, gold was often purified by fusion with antimony (p. 202), silver by cupellation (p. 204), and quicksilver by distillation.

The proximate materials of the Stone were then obtained as a result of further operations, which were preserved as sacred secrets by the adepts. These processes were sometimes represented as removing the earthy and watery principles from ordinary silver (or quick-

silver), which was then fixed by being deprived of its
airy or volatile principles: ' Separate the earth from the
fire, the subtle from the gross, acting prudently and with
judgment ', says the seventh precept of Hermes (p. 54).
The actual practical procedure [29] varied; but in conformity
with the maxim, *Until all be made water, perform no opera-*

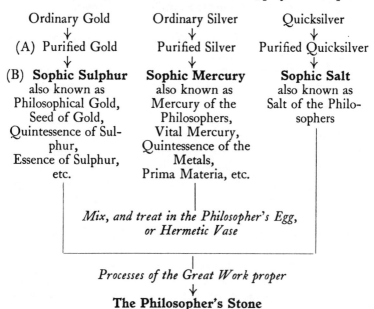

Ordinary Gold	Ordinary Silver	Quicksilver
↓	↓	↓
(A) Purified Gold	Purified Silver	Purified Quicksilver
↓	↓	↓
(B) **Sophic Sulphur**	**Sophic Mercury**	**Sophic Salt**
also known as	also known as	also known as
Philosophical Gold,	Mercury of the	Salt of the Philo-
Seed of Gold,	Philosophers,	sophers
Quintessence of Sul-	Vital Mercury,	
phur,	Quintessence of the	
Essence of Sulphur,	Metals,	
etc.	Prima Materia, etc.	

*Mix, and treat in the Philosopher's Egg,
or Hermetic Vase*

Processes of the Great Work proper
↓
The Philosopher's Stone

Fɪɢ. 11.—Diagrammatic Representation of a Simple Scheme for preparing the Philosopher's Stone: (A) Primitive, and (B) Proximate,
Materials of the Great Work.

tion (p. 262), the purified gold was usually dissolved
in *aqua regia* and the purified silver in *aqua fortis*. The
salts obtained upon crystallisation or evaporation were
then calcined. The resulting finely divided metals were
sometimes regarded as sophic sulphur and sophic mercury; sometimes, however, they were redissolved and
subjected to further operations. Sophic sulphur was
said to be fixed and red, sophic mercury to be volatile
and white. Together with a suitable form of sophic salt,

such as corrosive sublimate prepared from quicksilver and purified by sublimation, they formed the proximate ingredients of the Great Work.   As such, they were treated in the sealed Vase of Hermes, or Philosopher's Egg, according to an elaborate routine, described below (p. 136), which was held to lead to the Philosopher's Stone.

Gold was regarded as the best source of sophic sulphur, of which it was said to yield about half its weight.[30] Silver was supposed to contain a very pure form of mercury, of which ordinary quicksilver was an inferior source.   Plate 22 (i), showing the signs of sophic sulphur and sophic mercury coupled with those of gold and silver, respectively, indicates clearly the supposed relationships between these proximate and remote materials of the Stone.

The acids used in these processes were assigned fantastic names, such as ' vinegar of the mountains ' and ' stomach of the ostrich ' (capable of digesting anything!); they were usually symbolised as lions or other animals swallowing the sun and moon, or devouring serpents (Plate 38 (ii)).

Although the Philosopher's Stone was so often regarded as a triune microcosm (p. 110), the third principle, salt, does not always figure in its preparation.   In the scheme summarised in Fig. 11, salt may be envisaged as the medium uniting sulphur and mercury (p. 28); but sometimes the Stone was described, in terms of the original sulphur-mercury theory, as the result of the union of sophic sulphur and sophic mercury alone.   In still another conception of transmutation, sophic sulphur, or the seed of gold, seems to have been identified with the Stone, which when sown (p. 94) in a suitable menstruum, such as sophic mercury, was capable of converting it wholly into gold, thereby leading to a ' multiplication ' of the seed.   Thus, both the preparation of the Stone and its application as a transmuting agent could be regarded, according to these somewhat different views, as essentially the union of masculine and feminine principles.

The same fundamental idea led to the application of

such terms as the Hermetic Androgyne, and Rebis, to the Stone, or to the immediate result of the union of sophic sulphur and sophic mercury in its preparation. The Sun (sophic sulphur) was often termed the Father, and the Moon (sophic mercury) the Mother of the Philosopher's Stone. Occasionally the Stone was called the Royal Child. ' Its father is the sun, its mother the moon; the wind carries it in its belly, its nurse is the earth ', says the fourth precept of Hermes. Sometimes, also, as this reference shows, the elements of Aristotle were introduced into the scheme.

Alchemical writings on the Philosopher's Stone are concerned largely with accounts of the ' primitive materials '. These were not always identified with gold, silver, and quicksilver, as shown in Fig. 11. Some alchemists depicted the existence of a universal primitive matter, known as the Bird of Hermes, which was supposed to emulate the eye of its poetic creator by roving continuously from heaven to earth, and earth to heaven. Further, the imaginary Adamic earth, ' red earth ', or ' virgin earth ', was said—like the Stone itself—to be of universal occurrence; so that the secret of its identity had to be preserved by the esoteric brotherhood. ' When we have once obtained this,' wrote Isaac of Holland, ' the preparation of the Stone is only a labour fit for women, or child's play.' [31] Even this pronouncement, however, was possibly a play upon words. Women's work was sometimes held to be allusive to the feminine principle of the Great Work; but Trismosin [32] likens philosophical sublimation to ' Woman's Work, which consists in cooking and roasting until it is done '.

Among actual materials commonly used by seekers after the Stone were gold, silver, mercury, sulphur, salt, tin, lead, alum, quicklime, saltpetre, sal ammoniac, oil of vitriol, green vitriol, blue vitriol, and aqua vitae. The use of some of these was consonant with the dictates of alchemical theory. The unsophisticated ' puffers' used other materials in the operations, irrespective of any theoretical considerations; and it would be difficult to name any material known to them—whatever its nature, and however rare or repellent it may have been—which was not at some time or other drawn into the ambit of their operations.   Thus Chaucer [33] mentions,

> Arsenek, sal armoniak, and brimstoon.
> And herbes coude I telle eek many a one,
> As egrimoigne, valerian, and lunarie,
> And other suche, if that me list to tarie . . .
> Unslekked lyme, chalk, and glayre of an ey,
> Poudres dyvers and asshes, dong, and cley,
> Cerèd poketts, sal petre, vitriole;
> And dyvers fyres made of woode and cole;
> Salt tartre, alcaly, and salt preparat,
> And combust materes, and coagulat;
> Cley made with hors or mannes hair, and oyle
> Of tartre, alym, glas, barm, wort, and argoyle.

Equally comprehensive lists of materials are given by Thomas Norton (p. 178), Ripley,[34] Ben Jonson, and other writers.

In brief, although the details vary from one writer to another, the Philosopher's Stone was supposed to be attained by submitting suitable materials to an ill-defined series of practical operations, or processes.   The following summary, however, may be put forward as affording a general idea of the classification and sequence of the processes concerned:

(1) Purification of the primitive materials.
(2) Preparation of the proximate materials.
(3) Treatment in the Philosopher's Egg, or Hermetic Vase, with attendant colour changes.
(4) Increasing the potency of the resulting Stone (*multiplication*).
(5) Transmutation (in the operation of *projection*).

All of these operations were often comprehended in the expression Great Work, or Grand Magisterium; but some alchemical writers looked upon the first two kinds of processes as preliminaries to the Great Work proper.

PRELUDE TO CHEMISTRY

This distinction accounts partly for the divergencies occurring in the numerous descriptions of the number and sequence of the operations.

§ 8. INDIVIDUAL PROCESSES OF THE GREAT WORK

A general account of the processes adopted in purifying the primitive materials and preparing the proximate materials has already been given (p. 131). The remaining processes in the preparation of the Stone were usually depicted as occurring in the sealed vessel of Hermes, as a result of long-continued and carefully controlled heating (p. 144). Finally, the potency of the Stone was enhanced by means of a process known as 'multiplication', and the 'multiplied' Stone was then applied as a transmuting agent in the operation of 'projection'.

The number, names, and sequence of the practical operations in the Great Work vary greatly in the different accounts.* Pernety [35] quotes a list of twelve successive processes, which were suggestively represented by the signs of the Zodiac in the following manner:

1.	Calcination	♈	Aries, the Ram
2.	Congelation	♉	Taurus, the Bull
3.	Fixation	♊	Gemini, the Twins
4.	Solution	♋	Cancer, the Crab
5.	Digestion	♌	Leo, the Lion
6.	Distillation	♍	Virgo, the Virgin
7.	Sublimation	♎	Libra, the Scales
8.	Separation	♏	Scorpio, the Scorpion
9.	Ceration	♐	Sagittarius, the Archer
10.	Fermentation	♑	Capricornus, the Goat
11.	Multiplication	♒	Aquarius, the Water-carrier
12.	Projection	♓	Pisces, the Fishes

The *Twelve Gates* of Ripley [36] denote twelve processes also, and Ben Jonson in *The Alchemist* [37] gives a

* Among such accounts, that of Eirenæus Philalethes in the *Open Entrance to the Closed Palace of the King* [39] is at the same time picturesque and quasi-realistic.

136

PLATE 31

The 'Chymic Choir' of the Seven Metals.

From *Musaeum Hermeticum*, 1625. (See p. 167.)

PLATE 32

'Three Nurslings of the Wealthy Art'—Norton, Cremer and Basil
Valentine, with Vulcan.

From *Viridarium Chymicum*, Stolcius, 1624 (reprinted from the title-page of *Tripus
Aureus*, Maier, 1618).

PLATE 33

Elias Ashmole, F.R.S. (1617–1692).

From an engraving published by Weichard, London, in 1707. The symbol of Mercury refers to Ashmole's pseudonym *Mercuriophilus Anglicus*. (See p. 175.)

PLATE 34

A 15th-century Alchemist in his Laboratory.

From a 15th-century MS. copy of Norton's *Ordinall* in the British Museum (Add. MS. 10302, fol. 37 b). This is possibly the earliest representation of a balance in a case. The figure presumably represents Norton.

similar list of 'the vexations, and the martyrisations of metals in the work'. The series of twelve processes given by Mylius is enumerated below (p. 262). Norton, in one of his unprinted manuscripts,[38] gave a scheme of fourteen processes, arranged in the form of a so-called 'philosophic tree'. Paracelsus accepted the number of processes as seven. It was even held, by singularly optimistic adepts, that the Stone could be prepared from a single material, in a single vessel, at a single operation. In this, as in so many other respects, alchemy was generously latitudinarian.

Some of the names which were used to denote these processes are now obsolete. Others, such as solution (or dissolution), distillation, sublimation, calcination, and digestion, find a place in the vocabulary of modern chemistry, with little alteration of meaning. *Distillation*, however, was sometimes subdivided into *ascension*, or sublimation, and *descension*, or condensation; the separate processes were symbolised by a bird flying upwards or downwards, and the combined process by two birds opposing each other. Solution and fusion were both comprehended in the term *dissolution*; similarly, both crystallisation and solidification were denoted by *coagulation* or *congelation*, although the latter term usually implied the reverse of fusion. The adepts were not consistent in their use of terms; their language was apt to be vague; often, also, they attached a mystical significance to the materials, colours, processes, and apparatus of the Great Work. Bearing these points in mind, the nature of some of the processes may now be outlined.

Thus, in a treatise [40] attributed to 'Alphonso, King of Portugall', *solution* is described in the following terms: 'Our dissolution is no other thing, but that the body be turned againe into moistnesse, and his quicksilver into his owne nature be removed againe . . . the first worke of this worke is the body reduced into water; that is to Mercury, and that is that the Philosophers call solution, which is the foundation of all the worke'.

Although the royal alchemist regarded solution as the first operation in the Great Work (*cf.* p. 132), the preference was often given to *calcination*, or heating in open or closed vessels. The actual effect of this process was usually oxidation; but in the eyes of the alchemists the chief virtue of calcination was its power of ' fixing ' the more fusible metals, thereby depriving them of a property which rendered them unlike gold. A ' fixed ' metal was a metal which resisted change, and gold was regarded as perfect in this respect. *Separation* was an elastic term comprising such operations as filtration, decantation, and distillation of a liquid from suspended or dissolved matter: an alchemical definition of this process by Mylius is quoted elsewhere (p. 263).

The alchemists were particularly impressed by the phenomenon of *sublimation*. The deposition of crystals in the cool upper part of a vessel containing heated solid material in its lower part was likened by them to the upward flight of swans, doves, and other birds. The process was sometimes called ' exaltation ', or ' elevation ', and when repeated many times it was supposed to furnish the ' quintessence ' of the material concerned. In summarising the preparation of the Stone, Glauber [41] wrote: ' we may fitly subjoyn this sutable Poesie, making for our present Purpose, and expressing the same in few Words.

> ' *Dissolve the Fixt, and make the Fixed fly,*
> *The Flying fix, and then live happily.*'

Putrefaction, or *mortification*, was held to be necessary in order to enable the supposed seed of gold to germinate (p. 202), or to give ' another manner of being ' to the material of the Work. Fire was the common agent used in this process: the adepts held that while ' ordinary fire' destroyed the seeds of substances, the germinative power was unaffected by the action of various forms of ' philosophical fire '. The correlated process of *revivification*, or *resurrection*, was likened by the adepts to the restoration of the soul of matter to its body. The two changes were said to be marked by the appearance of black and white colours, respectively. Lead and other base metals were often regarded as dead metals, which would pass into silver or gold when revivified; the preparation of ' flowing mercury ' from cinnabar was also cited as an example of resurrection.

The process of *conjunction* was variously defined as the union, or marriage, of male and female, Sun and Moon, agent and patient, form and matter, sulphur and mercury, gross and subtle, fixed and volatile, brother and sister, Hermes' seal and wax, lion and serpent, toad and eagle, the two dragons, etc. (p. 103). ' The sperme of Sol is to be cast into the matrix of Mercury ', says Alphonso, King of Portugall.[40] These were double conjunctions. The union of body, soul and spirit, or the reduction of the trinity to unity, constituted the triple conjunction; and the quadruple conjunction was brought about by the union of the four elements.

Cibation consisted in supplying, or ' feeding ', the vessel with fresh material; it was sometimes likened to the nourishing of an infant with milk. *Cohobation* signi-fied the restoration of a distilled liquid to its residue, or fæces, followed by redistillation; this was an approach to the modern process of reflux distillation. The term *imbibition* was sometimes used in the same sense. Pernety defined cohobation as digestion and circulation of the material in the vase. *Circulation* was a primitive form of reflux distillation in a closed vessel, or double vessel.

Inhumation, according to Albertus Magnus,[42] was ' the placing of a soluble or dissolved substance in dung for purposes of dissolution '. *Ceration* implied bringing a material into a wax-like, soft, or fluid condition: ' test upon a hot plate, and if it melt swiftly like wax, the cera-tion is complete ', wrote Arnold of Villanova.[43] Some-times fermentation was used as a synonym for ceration.

Pernety defined *fermentation* proper as the rarefaction of a dense body by the interspersion of air in its pores;

other writers regarded it as an animating or resuscitating process. The Philosopher's Stone was often pictured as a ' ferment ' which could insinuate itself between the particles of imperfect metals, thereby attracting to itself all the particles of its own nature: thus the red Stone, or golden ferment, would yield gold, and the white Stone, or silvern ferment, would yield silver. ' It is incorrect to say that the alchemists seek to make gold ', wrote Pernety.[44] ' The first aim of the true philosophers is to find a remedy for the ills which afflict human nature; the second is to discover a ferment, which, when mixed with imperfect metals, is able to show that they contain gold, which, before the projection, was enclosed in them, in company with heterogeneous particles. . . . The ferment does no more than hasten a purification for which Nature requires ages, and which in some instances Nature could not effect at all in the absence of an active purifying agent.'

Similar ideas had been expressed by Salmon [45] about a hundred years earlier—if not, indeed, by Wei Po-yang in 142 A.D.!　It appears therefore that they were widely held. ' Lead is black on the outside but holds gold flower in its bosom ', wrote Wei Po-yang.[46]

Multiplication and projection, the last two processes of the Great Work, were immediately concerned in such ideas and claims as these. It was a fundamental tenet of alchemical doctrine that when the Stone, or ' powder of projection ', had once been made, it could be increased indefinitely in amount ; or, alternatively, that its power, or excess of perfection, could be increased by a process of concentration. This idea was embodied in the term ' multiplication '.*　It is not surprising that most writers use vague language in describing or discussing this process. Ripley [47] makes an illuminating comparison of the 'Medcyn ' with ' Fyer whych tyned wyll never dye: Dwellyng wyth the as Fyer doth in housys, Of whych one

---

* In everyday language the term ' multiplication ' was synonymous with ' transmutation ' or ' gold-making '.

sparke may make more Fyers I wys'. The Stone was also likened by other writers to a flint which could give fire to all and sundry without getting any smaller through it. Ripley adds, however, that the Medicine may be multiplied 'infynytly' with mercury; so that ten parts 'beyng multyplyed lykewys, Into ten thousand myllyons, that ys for to sey, Makyth so grete a number I wote not what yt ys'.

The general idea seems indeed to have been that the Stone could be multiplied by dissolving it in mercury, or in sophic mercury. There are also numerous references to a repetition of the operations of the Work with exalted and perfected materials. 'When do you make projection?' asks the impatient Mammon, in *The Alchemist*.[48] 'Son, be not hasty', admonishes Subtle, who has not yet multiplied his powder:

> I exalt our med'cine,
> By hanging him *in balneo vaporoso*,
> And giving him solution; then congeal him;
> And then dissolve him, then again congeal him:
> For look, how oft I iterate the work,
> So many times I add unto his virtue.
> As, if at first one ounce convert a hundred;
> After his second loose, he'll turn a thousand;
> His third solution, ten; his fourth, a hundred.
> After his fifth, a thousand thousand ounces
> Of any imperfect metal, into pure
> Silver and gold, in all examinations,
> As good as any of the natural mine.
> Get you your stuff here against afternoon,
> Your brass, your pewter, and your andirons.

'The oftener the Medicine is dissolved, sublimed, and coagulated, the more potent it becomes', wrote Arnold of Villanova; [49] 'in each sublimation its projective virtue is multiplied by ten'.

The actual transmuting process was called *projection*. This was the culminating operation of the Great Work. 'The red ferment has done his office', says Face to Mammon; [50] 'three hours hence prepare you to see projection.' In this final operation, the powder of projection, often enclosed in paper or wax, was thrown upon the hot quicksilver, molten lead, or other material to be transmuted. In some of the descriptions, a process of dilution, or of dilution and multiplication, is associated with the projection, owing to the extreme potency of the agent. Salmon [51] states that ' the Matter which transmutes, must be an hundred thousand times finer, more pure, and more exalted than the most fine Gold ': according to this view, the Stone was able to transmute owing to the excess of its perfection over that of natural gold.

In other accounts, the imaginary Stone seems to have been visualised as an agent capable of initiating a process of development of base metals into gold; and here, the properties ascribed to it are distinctly reminiscent of the behaviour of enzymes, or of the equally potent and astonishing hormones of animals and plants.

The perfect Stone, according to the pseudo-Roger Bacon, could transmute a million times its weight of a base metal into gold. The pseudo-Lully raised the proportion to a thousand billion parts, and indulged in the exuberant but perfectly safe boast that he would transmute the sea into gold if only it consisted of mercury! He did not attempt to explain how such an achievement would benefit himself or others. Philalethes, in his *Guide to the Celestial Ruby*,[52] wrote that ' your arithmetic will fail sooner than its all-prevailing power '. According to the mysterious Basil Valentine, however, the amount was limited to seventy parts; and the ill-fated Dr. John Price, a late transmutationist of the eighteenth century, more modest still, claimed only a thirtyfold transmutation.

Some adepts held the view that artificial gold—which significantly was usually described as ' better than gold from the mines '—inherited a transmutative power from the Stone which had been used in its production, owing presumably to the presence among its particles of some of the unexpended ' ferment ' (p. 206).

THE PHILOSOPHER'S STONE

§ 9. THE RÔLE OF FIRE

The adept's anxieties did not end with a proper selection and arrangement of his processes; for this exacting Great Work demanded also a careful control of other conditions. Among these, particular importance was attached to the mode of heating, the nature of the vessels used, the astrological influences, and the duration of the Work. Besides all this, many of the adepts held that an animated heavenly breath, or similar influence, was necessary for the completion of the Work—possibly to effect the union of ' the powers of things superior and inferior ', indicated in the eighth precept of Hermes— and that the worker should be endowed with certain moral and spiritual qualities (p. 178). Truly, the adept's lot was neither happy nor easy.

Fire was the most important agent of the alchemists, and their accounts of its varieties and powers are legion. Emblematically, it is denoted by various cutting or wounding implements, such as scissors, sword, lance, scythe, arrow, or hammer. It is symbolised by Vulcan, or Hephaestus, the god of fire, who, having been born lame, is often depicted as a man with a wooden leg. A similar symbol is used for Saturn, or Kronos (lead), but in this case the man is aged and carries a scythe, or sickle, and sometimes an hour-glass (p. 62).

In many writings dealing with the operations of the Great Work, much capital is made of a secret and mysterious Fire of the Sages, or Fire of the Philosophers. This is described as a celestial fire, creative in effect, and quite distinct from common or ' elemental ' fire, which is characterised as brutal and destructive.

Fire (or heat) was graded in regimens, or degrees, corresponding usually to different temperatures. Fire of the first degree was sometimes defined as equal to that of a brooding hen. Other adepts called the first degree the ' fire of Egypt ', holding it as equivalent to the Egyptian summer temperature. As the Great Work progressed, the temperature was raised through the second and third to the fourth degree, approximating to the melting point of lead. According to another scale, the Sun in Aries, Leo, and Sagittarius indicated the first, second, and third grades of the Fire (*cf.* Plate 23 (i)). Different degrees, or kinds, of heat (fire) were denoted also as water-heat, sand-heat, and naked-heat: these were achieved by using the water-bath, sand- or ash-bath, and bare flame.

Ordinary fire, or natural innate fire, was usually viewed by the alchemists as a liquid constituent of matter, of a viscous or oily nature—the equivalent of ' sulphur ' of the early theory and ' phlogiston ' of a later day. The fire (heat) derived from the sun was assumed to be more refined than terrestrial fire. In general, three species of fire were recognised, namely: celestial, subterraneous, and culinary.

' A parfet *Master* ye maie him call trowe, Which knoweth his Heates high and lowe.' The writings of Thomas Norton (p. 181), among others, convey a striking impression of the importance which the alchemists attached to the use and control of fire in their experiments. ' Nothing ', he states,[53] ' maie let more your desires, Than ignorance of Heates of your Fiers '; and he goes on to say that many of ' Gebars Cookes ' were deceived in this matter, despite their great knowledge of books. He talks of dry and moist fires, and of ' fiers ', or ' heates ', of ' Disiccation ', ' Conservation ', ' Coaction ', ' Calcination ', and so forth. Some alchemists, indeed, attached so much importance to the control of temperature that they used oil-lamps, with adjustable wicks of different sizes, as the source of heat. The fire, once kindled, was maintained without interruption until the consummation of the Work.

' Alphonso, King of Portugall ' observes [40] that ' the Philosophers in their books have chiefly put two fires, a dry, and a moyst; for the dry fire, they call it the common fire, of any manner of thing combustible that will burne: but the moist fire they call the hot, *venter Equinus*, which may be Englished, the Horse belly; but rather it is Horse dung, wherein remaining moystness, there doth remain heat.'

' I must have more coals laid in ', says Subtle, in *The Alchemist*,[54]

> We must now increase
> Our fire to *ignis ardens*, we are past
> *Fimus equinus, balnei, cineris,*
> And all those lenter heats.

The 'lenter' heats here enumerated by Ben Jonson (Plate 24) include those of horse dung, the water-bath, and bath of ashes or sand.

The Chinese alchemists, also, paid great attention to their fire. Ko Hung [55] wrote, in the fourth century A.D., that in compounding the gold medicine ' the fires should be tended for tens of days and nights with industrious application and close adjustment, which is a great difficulty '. An interesting passage in the *Ts'an T'ung Ch'i* of Wei Po-yang,[56] of the second century A.D., runs as follows: ' Treatment and mixing will bring about combination and rapid entrance to the scarlet portal. The escape must be firmly blocked. Below plays the dazzling flame, while the Dragon and Tiger keep up a sustained vociferation. The flame at the start should be weak, so as to be controllable, and should be strong at the end. Close attention and careful watch should be given so as to regulate properly the heat and cold. There are twelve divisions to the cycle. On the completion of the cycle a closer watch should be accorded. As the breath expires, life is ended. Death expels the spirit. The colour changes into a purple. Behold the *huan-tan* (Returned Medicine) is obtained.'

§ 10. THE SEQUENCE OF COLOURS

The supposed connection between the degree of heat and the progress of the Great Work was naturally linked up also with colour changes. ' The Matter of the Sages ', wrote Basil Valentine in his Ninth Key (p. 203), ' passes through the several varieties of colour, and may be said to change its appearance as often as a new gate of entrance is opened to the fire '. A quaint account of the successive changes in colour supposed to be brought about by continued heating in ' Our Furnace called the *Philosophers Dunghill* ' is given in *Bloomefields Blossoms*.[57]

Reference has been made in an earlier chapter (p. 13) to the importance which the Alexandrian makers of imitation gold and silver attached to colour changes occurring in alchemical processes. A like insistence upon the primary significance of colour, and colour sequence, permeated the alchemy of the Middle Ages and the succeeding centuries. The various stages necessary for the accomplishment of the Great Work, or preparation of the Philosopher's Stone, were supposed to be marked by the appearance of characteristic colours; and unless these colours followed one another in a proper sequence the operations were dismissed as useless.

The principal colours were said to develop in the order, black, white, citrine, and red; these colours were sometimes associated, respectively, with earth and black bile, water and phlegm, air and yellow bile, fire and blood —that is, with the four elements and the four humours. Thus, Paracelsus [58] states: ' Soon after your Lili [tincture] shall have become heated in the Philosophic Egg, it becomes, with wonderful appearances, blacker than the crow; afterwards, in succession of time, whiter than the swan; and at last passing through a yellow colour, it turns out more red than any blood '.

Of various subsidiary colours, grey followed black, green followed white, and the rainbow colours of the peacock's tail also appeared during the process: ' Betwixt Black and Whyte sartayne, The Pekokes fethers wyll appeare plaine ', wrote Thomas Charnock [59] in 1574. Philalethes, writing in 1645, gave the sequence as black, the peacock's tail, white, orange, and red.[60]

Basil Valentine and others denoted the black, white, rainbow, and red colours by symbolic representations of the crow (or raven), swan, peacock, and phoenix, respectively. The black colour was held to signify complete putrefaction and solution—or, according to some writers, complete conjunction—of the material; white marked its partial fixation, and red its complete fixation and perfection. ' Red is last in work of *Alkimy* ', wrote Thomas Norton (p. 179) in 1477. More than thirteen hundred years earlier, Wei Po-yang [61] had written: ' the Red Bird is the spirit of fire '.

The appearance of red before black showed that the material had been overheated, and the Work had then to be started afresh. Fire was denoted by such terms as sword or scissors (p. 143): thus, to cut off the head of the Black Crow meant to continue the heating, digestion, etc., until the black colour changed to white. If a black colour put in an unwanted appearance later in the process, the alchemist exclaimed irascibly to his ' yeoman ' or ' minister ' that the young ones of the crow were going back to their nest—and then, ' there never was such woe or anger or ire' [62] in the laboratory. All the colours of the Great Work were supposed to reappear, in a more rapid and transitory manner, during the operation of multiplication. Sometimes the stages of the Work denoted by the appearance of characteristic colours were known as regimens: such were the regimen of Saturn (black), the regimen of the Moon (white), the regimen of Venus (green and purple), the regimen of Mars (rainbow), and the regimen of the Sun (red or golden). These regimens were sometimes represented by flowers.

In the Recapitulation of his *Twelve Gates*, Ripley [63] refers to the colours of the Great Work in the following terms:

Pale, and Black, wyth falce Citryne, unparfyt Whyte & Red,
Pekoks fethers in color gay, the Raynbow whych shall overgoe
The Spottyd Panther wyth the Lyon greene, the Crowys byll bloe
    as lede;
These shall appere before the parfyt Whyte, & many other moe
Colors, and after the parfyt Whyt, Grey, and falce Citrine also:
    And after all thys shall appere the blod Red invaryable,
    Then hast thou a Medcyn of the thyrd order of hys owne kynde
    Multyplycable.

The emphasis attached to the production of colours in the preparation of the Stone was in keeping with the conception of its function as a dyeing or ' tingeing ' agent for the imperfect metals: because of this belief, the Stone was often known as the Red Tincture. ' Mare tingerem, si mercurius esset! ' claimed the pseudo-Lully, in the grandiloquent manner of the alchemists: ' I would tinge the sea if it were made of mercury! ' ' Gold tinges other metals to Sol ', wrote Paracelsus.[64] According to this view, gold, the perfect metal, has a yellow colour which is permanently ' fixed ', since it resists the action of chemical agents and of fire. Brass is therefore regarded as copper on the way to becoming gold; but since the yellow colour is unable to resist certain agents, the applied tincture, or tinctorial agent, is incomplete, and the transmutation is imperfect. Similarly, the temporary whitening of copper by mercury denotes an imperfect transmutation of copper into silver.

Thus, an epigram attached to an engraving (Plate 23 (ii)) in the *Viridarium Chymicum* (1624) runs as follows: ' A woman sometimes mixes various colours, and straightway washes therein linen or clothes. But the water departs and evaporates into thin air; the linen remains dyed with the desired colour. So the Water of the Sages penetrates the members of the metals, and in its swift flight makes bodies coloured.' Ripley expresses the same idea in verse: [65]

> Which Tinctures when they by craft are made parfite,
> So dieth Mettalls with Colours evermore permanent,
> After the qualitie of the Medycine Red or White;
> That never away by eny Fire, will be brente.

§ 11. THE VASE OF HERMES

It is a further striking feature of alchemy that the forms of the vessels used in the practical operations were supposed to exert a mystical influence upon the character of the product. The form of the ' double pelican ', for

example (Fig. 12), was mystically connected with the
process of conjunction.   Air, water, and earth were recognised by certain adepts as the vessels in which Nature
conducted her operations.   The alchemical vessel *par
excellence* was the so-called Aludel, Hermetic Vase, Vase
of the Philosophers, or Philosopher's Egg, in which the
Great Work was consummated.   This was shaped like
an egg, and thus resembled the symbol of creation of the
ancients.   It is often depicted (Plates 32 and 35) with an

FIG. 12.—The Double Pelican.
From the *Buch zu Distillieren*, Brunschwick, 1519.

enclosed serpent, symbolising the material of the Philosopher's Stone.

In the words of the informative ' Alphonso, King of
Portugall ', ' the forme of the glasse must be of the forme
of the Sphere, with a long neck, and no thicker than can
be grasped with a large hand, and the length of the necke
not above a span, and no wider then the Ægyptian seale
may cover its mouth . . . even as there be in a naturall
Egge three things (*viz.*) the Shell, the White, the Yolke;
even so there be in the Philosophers Stone three things,
(*viz.*) the Vessell, the Glasse, for the Egge shell, the white

liquor for the White of the Egge, and the yellow body for the Yolke of the Egge; and there becomes a Bird of the yellow and white of the Egge, by a little heate of the Mother, the Egge-shell still remaining whole untill the Chicken doe come forth; even so by every manner of wise in the Philosophers Stone, is made of the yellow body, and white Liquor by mediation of a temperate heat of the mother the earthly substance *Hermes* bird, the vessell still remaining whole, and never opened untill his full perfection.'

Common forms of the Vase or Egg are represented repeatedly in alchemical illustrations (Plates 27 and 35). The capacity was laid down by Philalethes as ' twenty-four full Florence glasses, neither more nor less '. He says further,[66] ' let the height of the vessel's neck be about one palm, and let the glass be clear and thick . . . in order to prevent the vapours which arise from our embryo bursting the vessel '. Since the vessel was ' hermetically sealed ' in the operations of the Great Work, it was sometimes called the House of Glass, or Prison of the King. The ' little mountains ' was another term applied to the Vase and its containing furnace; the ' white herb growing on the little mountains ' was thus the white stage of the magistery. Here is a description from ancient China: ' Like the Moon lying on her back is the shape of the Furnace and the Pot ', wrote Wei Po-yang.[67] ' In it is heated the White Tiger.' The various ways of heating the Vase or Egg have already been mentioned (p. 143).

In the version of Trismosin's *Splendor Solis* (p. 67) contained in the Harley manuscript,[68] there are seven very beautiful paintings of the Hermetic Vase, depicting successive operations in the Great Work. In each instance the Vase is surmounted by a crown and placed within an ornamented niche. This brightly coloured subject is bordered by a subsidiary design, delicately drawn and tinted, in which various aspects of mediaeval life and activities are treated with much skill and charm. At the top of each painting is an allegorical device of astrological import. The seven Vases contain, respectively: a naked child pouring a liquid down the throat of a dragon; a red and white bird pecking a black one; a white bird with three crowned heads; a dragon with red, white, and black heads; a gorgeously coloured peacock; a queen wearing a pale-blue robe ' typifying the mystical sea '; [69] a king attired in red and gold, holding an orb and sceptre, and standing upon a crescent moon.

These paintings are allusive, *inter alia*, to the changes in colour which were supposed to mark the alchemical operations. The fifth painting which is in the British Museum, depicts a peacock enclosed within the Hermetic Vase. Above, Venus' car, drawn by two doves, is marked with the signs of the Ram and Scales (Aries and Libra). Below, several musicians in brightly coloured attire are playing and singing: the seat occupied by the guitar-player bears the date of the manuscript (1582). In addition, the border shows a group of people seated at table; numerous bathers in a stream; and couples engaged in walking, dancing, riding, and love-making.

The sixth painting of the Vase, showing the advent of the Queen, bears the astrological sign of Mercury in Gemini and Virgo. In the background of the border are depicted the streets of a busy mediaeval city. In the foreground, Sculpture is represented by two men with mallet and compasses; Mathematics by two philosophers with a globe and compasses; Literature by an old scholar seated at a desk; Commerce by a man counting money; and Music by a little company of musicians and singers grouped around an organ. The last painting of this series denotes the consummation of the Great Work, and carries the sign of the Moon in Scorpio (p. 152); its border illustrates the sports of fishing, shooting, hawking, and hunting.

A remarkably elaborate treatment of the processes which were supposed to occur within the Vase of Hermes is given in a series of seventy-eight symbolic illustrations in the third and last part of J. C. Barchusen's *Elementa*

Chemiae, published at Leyden in 1718. The accompanying comments contain numerous references to alchemical literature, in particular to the *Rosarium Philosophorum* (p. 318). The seventy-sixth picture shows the Vase enclosing the crowned King, symbolising the achievement of the Great Work; the seventy-seventh depicts an angel unrolling a scroll of music (p. 254), bearing the words *Gloria Laus et honor Deo in excelsis*; and the seventy-eighth, and last, is clearly intended to express the same idea as the last picture of the enigmatical *Mutus Liber* (p. 159), although the two designs are quite distinct.

Certain alchemical writers sought to maintain their reputation for mystic expression by stating that the Vase of the Philosophers was their Water, and that the Vase and its containing Furnace were one and the same thing. Others, however, have given descriptions, and even drawings (Plate 22 (ii)), of the Philosophical Furnace, or Athanor, the identity of which was often cloaked under such fanciful names as Sepulchre, Triple Vessel, Green Lion, and House of the Chick. As the last name indicates, it was regarded by some of the adepts as an incubatory or brooding furnace. The Philosophic Furnace and the enclosed Vase are depicted repeatedly in the *Mutus Liber* (Plate 27). An ancient Chinese alchemical furnace, or *ting*, is shown in Plate 20.

§ 12. INFLUENCE OF THE STARS

As mentioned above (p. 150), each of the seven paintings of the Hermetic Vase contained in *Splendor Solis* bears a prominent astrological device: this detail illustrates the alchemical belief that in order to achieve the Stone it was necessary to conduct each operation under the proper astrological conditions.

Speaking of the ' elements of the Stone ', Norton emphasises that ' thei being in warke of Generation, Have most Obedience to Constellation '; moreover, he gives

PLATE 35

Basil Valentine.

Matth. Scheits *del.*, Melchior Hasner *sc.*, Anno 1677. The frontispiece to *Fr. Basilii Valentini Benedictiner Ordens Chymische Schriften*, Hamburg, 1677. The emblem of the Third Key is shown in the background.

PLATE 36

The Triumphal Chariot of Antimony.

Ramyn de Hooghe *fecit*, 1674. The frontispiece to
*Theodori Kerckringii doctoris medici Commentarius in
Currum Triumphalem Antimonii Basilii Valentini*,
Amsterdam, 1685.

PLATE 37

(i) The First Key of Basil Valentine.

From *Viridarium Chymicum*, Stolcius, 1624 (reprinted from *Tripus Aureus*, Maier, 1618). (See p. 200.)

(ii) The Eighth Key of Basil Valentine.

From the same.

PLATE 38

(i) The Ninth Key of Basil Valentine.

From *Viridarium Chymicum*, Stolcius, 1624 (reprinted from *Tripus Aureus*, Maier, 1618).

(ii) The Twelfth Key of Basil Valentine.

From the same. (See p. 205.)

some specific astrological diagrams (p. 100), in which
' The Planets vertue is proper and speciall ' for the Work.
It is significant that the operations of the Work were
sometimes denoted by the signs of the Zodiac (p. 136),
being made twelve in number for that purpose.   The
operations were even conducted by some adepts in a so-
called ' cosmic furnace ', which was depicted as a microcosm with its own Zodiac, seasons, and poles.[70]
    ' Let me see ', ruminates Subtle in *The Alchemist*,[71]

> ' How is the moon now? eight, nine, ten days hence,
> He will be silver potate; then three days
> Before he citronise: some fifteen days
> The magisterium will be perfected.'

§ 13. DURATION OF THE GREAT WORK

Astrological conceptions are in turn closely bound up
with the many time-tables set forth by seekers after the
Stone.   The duration of the Great Work varies greatly
in the different descriptions.   Some of the adepts likened
their Work to the creation of the world in seven days; for
this reason, and also because of an imagined mystical
relationship with the seven metals and the seven planets,
they adopted a scheme of seven operations occupying
seven days.

    Such hasty methods, however, found no favour among
the majority of the cognoscenti, whose views squared
rather with those of the Carpenter in Lewis Carroll's
immortal poem: [72]

> ' If seven maids with seven mops
>   Swept it for half a year,
> Do you suppose,' the Walrus said,
>   ' That they could get it clear? '
> ' I doubt it ', said the Carpenter,
>   And shed a bitter tear.

Norton states that he ' learned all the secreats of
Alkimy ' in a philosopher's month of forty days.   Accord-

ing to Ripley, a year was necessary for the complete fruition of the operations. ' If you operate on Mercury and pure common gold,' writes Philalethes,[73] ' you may find " our gold " in 7 to 9 months, and " our silver " in 5 months. But when you have these, you have not yet prepared our Stone: *that* glorious sight will not gladden your eyes until you have been at work for a year-and-a-half.' He adds, rather unnecessarily, that patience is the great cardinal virtue in alchemy.

Other estimates were much longer still, running to as much as three, seven, or twelve years. Some alchemists, however, adopted an alchemical time-scale: the four main colours of the Work sometimes denoted the four seasons, and the time occupied by each operation was called a year. Nevertheless, the tedious nature of the processes is frequently expressed. It is stated, for example, that the black colour, or regimen of Saturn, may last from the fortieth to the ninetieth day. A poem in Ashmole's collection [74] lays down that—

> The *Glasse* with the Medicine must stand in the fyre
> Forty dayes till it be Blacke in sight;
> Forty dayes in the Blacknesse to stand he will desire,
> And then forty dayes more, till itt be White,
> And thirty in the drying if thou list to doe right.

§ 14. ENIGMAS OF THE STONE

Alchemical literature abounds in enigmatical drawings, diagrams, formulae, and other cryptic representations of the secrets of the Great Work. One of the earliest representations of this kind is the mysterious Formula of the Crab (Fig. 4, p. 40), in which Zosimos the Panopolitan is said to have concealed the secret of transmutation in the fourth century. The Formula conveys, to the initiated, a method of colouring base metals, by the use of copper compounds, so as to make them resemble gold. ' In this formula', says Campbell Brown,[75] ' we have the last remnant of an ancient symbolism, that was quite meaningless without *viva voce* exposition. The method of working was in this manner handed down from one ·generation of craftsmen to another.'

Of the later enigmas purporting to give directions for the achievement of the Stone, some of the best known occur in the *Twelve Keys* and other writings ascribed to Basil Valentine; these are discussed elsewhere (pp. 200, 208). Several of them carry the famous ' vitriol ' acrostic in the design. The fullest form of this was as follows, but the last two words were often omitted: *Visitabis interiora terrae, rectificando invenies occultum lapidem, veram medicinam.* The initials thus gave the word *Vitriolum.* Sometimes the words *occultum lapidem* were replaced by *oleum limpidum.* ' Visit the inward parts of earth; by rectifying thou shalt find the hidden Stone, the true Medicine ': some of the adepts held that this formula concealed the whole secret of the Great Work and its material; but they differed greatly among themselves in their interpretation of the word ' vitriol ', which was a generic term applied to a large variety of substances (p. 309).

There is a very remarkable pictorial addition to the first volume of Manget's *Bibliotheca Chemica Curiosa*, of 1702 (p. 116). The supplement is unmentioned in the text of this work, and it carries no description beyond the title.* It consists of fifteen engraved plates of folio size (about 19 by 27 cm.), the first of which bears the words: *Mutus Liber, in quo tamen tota Philosophia hermetica, figuris hieroglyphicis depingitur, ter optimo maximo Deo misericordi consecratus, solisque filiis artis dedicatus, authore cuius nomen est Altus* (' The Wordless Book, in which nevertheless the whole of Hermetic Philosophy is set forth in hieroglyphic figures, sacred to God the merciful, thrice best and greatest, and dedicated to the sons of art only, the name of the author being Altus').

This work is one of the outstanding enigmas of al-

---

* The cryptic references on the first plate, reversed in the Hebrew fashion, denote *Genesis*, 28, 11, 12, 27, 28, 39 ; *Deuteronomy*, 33, 19, 28 [*cf.* also 13, 14].

chemical literature, and as it deals in a unique way with the operations of the Grand Magistery it merits more than passing notice. Its authorship is uncertain, and no satisfactory interpretation of its contents has ever been published. Ferguson [76] remarks that 'the plates are partly symbolical, partly pictorial, representing an alchemist and his wife engaged in chemical operations, such as sublimation, distillation, and the hermetic sealing of flasks. . . . There is hardly a clue, however, to the substances symbolised.' The engravings were first published at La Rochelle, by Pierre Savouret, in 1677. Arcere,[77] a local historian, writing in 1757, sought to father this mysterious *Livre Muet* upon one Saulat, Sieur des Marez; but others have ascribed it to a physician and alchemist of La Rochelle, named Tollé, who was also an artist. A point apparently neglected is that the first engraving shows a young man sleeping, with a stone for his pillow. He lies at the foot of a ladder reaching from earth to heaven: and angels blowing trumpets are ascending and descending on it. The young man may therefore be identified as Jacob;[78] possibly, also, the stone that Jacob 'put for his pillow', and upon which he poured oil, was accepted as a symbol of the Philosopher's Stone.* Now Jacob was the Christian name of Saulat.

There is little doubt that the pseudonym 'Altus' may be identified with the 'Elder', 'Senior', or 'Vieillard', of alchemical literature (pp. 268, 276). These names may refer to the Ancient of Days; they were often applied to sophic sulphur, and less frequently to sophic mercury. Alternatively, Altus may be Flamel's 'great old man' with the hour-glass (p. 62), representing Saturn, or lead. The Elder, 'who brings the key' (p. 272), is one of the chief actors in the series of episodes here portrayed; the others

---

* *Cf.* Ashmole's reference to Jacob's ladder (p. 92). An early 18th-century MS. in the St. Andrews collection, entitled *Liber Mutus* (p. 158), refers to 'Denis Tollé médecin à la Rochelle' and states that his friend 'le S^r jacob Saulat Desmarets', the author of the work, anagrammatised his name to 'Altus'. De Monconys (*Voyages*, 1677, 1, 20) described 'M. Tole le Medecin', in 1645, as a chymist and artist.

are the adept and his wife, Mercury, Sol, Luna, and, lastly, an infant, representing the Philosopher's Stone.

In the fourth plate (Plate 25) the scene is laid in a field, in which the adept and his wife are wringing dew out of a cloth, seemingly under a heavenly influence and subject to the astrological control of the Sun in Aries and the Moon in Taurus. This proceeding is reminiscent of a passage in a contemporary work by Salmon,[79] which opens with the direction: 'Gather Dew in the Month of *May*, with a clean white Linnen Cloth spread upon the Grass'. When this filtered May-dew is digested for fourteen days in horse-dung, and then distilled to a quarter of its bulk four times running, it is said to yield a potent Spirit of Dew; 'and if you are indeed an Artist, you may by this turn all Metals into their first matter' and prepare 'an Elixir of a wonderful virtue in Transmuting of Metals'.* Dew was sometimes identified with the *Spiritus mundi*, or 'spirit of the world', a hypothetical spirit or material which the alchemists endowed with many marvellous properties, including the power of dissolving gold.

In the fifth plate, the dew—which is perhaps to be interpreted as dew of Hermes, rain of the Sages, or celestial water (p. 103)—is distilled; the still-residue is taken by a man marked with the sign of Luna, and carrying an infant, which here makes its *début*; the distillate is heated in a furnace marked '40', indicating possibly a period of forty days.

The sixth plate illustrates more distilling operations, and a flower—possibly the *Sanguis Agni* [80]—is shown in the still, indicating the appearance of a crucial colour; the adept hands the residue to a figure of a man, who wears a tunic and carries a bow. In the seventh engraving there is more heating in an open dish, and the wife transfers the residue to a bottle marked with four stars; the

---

* According to another account, May-dew, when distilled with *aqua fortis*, mixed with sublimated mercury, and putrefied for a month in warm horse-dung (*fimus equinis*), was said to yield the wonder-working 'virgin's milk' (*lac virginis*).

Elder, with the infant in his arms—both being naked—sits on a bonfire; he is then depicted in a tub, still carrying the infant, and undergoing anointment with one of the products of distillation; later, the infant is held by a man with a scimitar, while Luna stands by with the bottle marked with four stars; this man and woman (Luna) appear to be the adept and his wife, although the figures are now unclothed (p. 221). Here, the bonfire suggests calcination; the tub, ablution; the scimitar, treatment with fire; and the four stars, sophic gold, or the *materia prima* of the Stone. Alternatively, the man with the scimitar may be a seventeenth-century reflection of Flamel's ' *King* with a great *Fauchion*, who made to be killed in his presence by some *Souldiers* a great multitude of little *Infants* ' (p. 62). Moreover, in two of the drawings the Elder seems to be biting the infant's hand. Since ' infant's blood ' was an expression for the ' mineral spirit of metals ', and the Elder sometimes represented sophic sulphur (p. 272), this curious detail may signify the conjunction of sophic sulphur and sophic mercury. Or, it may represent Saturn (Kronos) swallowing his offspring (p. 243): in this way, lead, symbolised by Saturn, may be pictured as undergoing transmutation into gold through the intervention of the ' infant's blood '. The St. Andrews collection has a complete set of the drawings, with a short manuscript in French entitled *Liber Mutus* (p. 156). The manuscript refers to Manget's publication, and the drawings are apparently a refined French version of the original designs of 1677. Here, in the seventh plate, the Elder unmistakably bites the infant's elbow and hand (Plate 26).

The tenth plate marks the decisive operation. Equal weights of two liquids, marked with a star and a flower, possibly symbolising the proximate materials of the Great Work (p. 131), are poured into an aludel, which the adept seals hermetically with a blowpipe; the vessel, which has now become the sealed Vase of Hermes, or Philosophic Egg, is placed in the Athanor, beside which stand two figures marked with the signs of Sol and Luna, symbolising sophic sulphur and sophic mercury. Luna carries a bow, and her companion (Sol) is the man in a tunic who had the bow in the sixth plate; on the other side of the furnace a target is shown, divided into four concentric segments; the numeral ' 10 ' is marked at the foot of each figure. Here, in the Furnace of the Sages, the Great Work is consummated: the adept and his wife—who are suggestive of the masculine and feminine principles represented by ' the *Red Man* and hys *Whyte Wyfe* ' (p. 101)—have ' hitt the *Marke* ' (p. 130) and achieved the Stone.

The next three plates evidently symbolise multiplication of the Stone which was achieved in the tenth plate. In the thirteenth plate (Plate 27) the flower is replaced by the symbol of Sol, and the number ' 10 ' is replaced by ' 100, 1000, 10000, etc.' In the fourteenth plate the adept and his wife open the Philosophic Egg, and, in the lower part of the engraving (page 278), return thanks to God and take a vow of silence. Between them appears the alchemical aphorism: ' *Ora Lege Lege Lege Relege Labora et Invenies* (' Pray, read, read, read, again read, toil; and thou shalt find ').

The final engraving has the same distinctive setting as the first; the ladder is down, and an aged man is being carried up to heaven by two angels; a figure (probably Hercules) lies extended upon a lion's skin; the symbols of Sol and Luna are shown near by; from the mouths of the kneeling adept and his wife rise scrolls with the inscription *Oculatus abis* (' You depart seeing ').

The *Mutus Liber* undoubtedly depicts the operations of the Great Work. Moreover, by its form, the author seems to suggest that, while a young man, he was vouchsafed a heavenly vision of these operations; after ' long labour unto aged breath ', he has succeeded in his quest; his eyes having seen the earthly crown, he departs to seek a still greater reward in heaven. The last scene is, indeed, the ' Nunc Dimittis ' of Altus.

'When the aspirant is accomplished,' wrote Wei Po-yang,[81] 'he will ride on the white crane and the scaled dragon to pay respects to the Immortal Ruler in the Supreme Void. There he will be given the decorated diploma which entitles him to the name of a *Chên-jên* (True man).'

§ 15. MYTHOLOGY AND THE STONE

The main trend of the meaning of the *Mutus Liber* is clear; but there are puzzling details, in particular the ascription of the symbol of Luna both to male and female figures, and the transference of the bow from a male to a female. It is possible that the bow is allusive to Apollo, the expert archer, and Artemis (Diana), who would typify the relationship of brother and sister, to which some of the alchemists attached significance (p. 103). Diana and Apollo were sometimes used as synonyms for the white and red colours of the Magistery; Philalethes called the white colour 'Diana's doves'. It is also of interest that the story of Cupid and Psyche was associated with the operations proceeding in the Vase of Hermes: thus, the precipitate departure of Cupid, brought about by a drop of 'incombustible oil' falling upon his shoulder from Psyche's lamp, was held to be an image of the volatilisation of material from the lower to the upper part of the Vase.

The repeated allusion to classical mythology is, indeed, a feature of the literature of the Philosopher's Stone which calls for particular comment. Thus, Suidas held that the Golden Fleece of the Argonauts was a papyrus containing the secret of gold-making. Much later in the history of alchemy, Michael Maier (p. 234) emphasised the supposed connection between alchemy and mythology; finally, in the middle of the eighteenth century, Pernety (p. 245) developed the startling thesis that classical mythology from the beginning was subservient to alchemy, and that it was conceived expressly to record and conceal the sacred truths of the Hermetic doctrines.

THE PHILOSOPHER'S STONE

Thus, some of the adepts, with considerable justification, compared the operations of the Great Work with
the labours of Hercules; further, the alleged misguided
efforts of the ' puffers ' were ironically correlated with
the unending task of Sisyphus. Ben Jonson [82] links up
the latter conceit very neatly with the alchemical craze for
obscurity of expression:

> Was not all the knowledge
> Of the Egyptians writ in mystic symbols?
> Speak not the Scriptures oft in parables?
> Are not the choicest fables of the poets,
> That were the fountains and first springs of wisdom,
> Wrapped in perplexed allegories?
> I urged that,
> And cleared to him that Sisyphus was damned
> To roll the ceaseless stone, only because
> He would have made ours common.

The puffers are sometimes likened to Penelope, who undid
by night what she did by day; the Odyssey is an exposure
of the errors they make at every step they take; Ulysses
is the true adept who overturns their ill-conceived pro-
jects. The serpent-haired Medusa, by virtue of her spectacular power, is naturally linked with other fevered
conceits concerning the Philosopher's Stone.

Albification, or whitening of the black putrefactive
stage of the Great Work, is often bound up in alchemical
writings with veiled references to Leto, Lato, or Latona,
the Titaness. In Greek mythology, Latona aroused the
jealousy of Juno by responding to the amorous advances
of Jupiter (Zeus): her twin children, Apollo and Artemis
(Diana), were symbolical in alchemy of Sun and Moon, or
masculine and feminine principles. Python, the serpent
with which Juno sought to terrorise Latona (p. 165),
and which was eventually slain by Apollo, also figures
repeatedly in alchemical lore. Laton, or Latona, the
material of the Great Work at the stage of putrefaction
or darkness—also known as *Saturne, our Brasse,* the
Phylosophers Lead, or the *discontinuous powder* [83]—was said

to be whitened by azoth or sophic mercury. It was then called whitened laton, and the success of the operation was assured. By sowing or implanting a red colour in whitened laton, the red stage of the Work was achieved. White and red laton thus correspond to the Queen and King, or Luna and Sol, in these operations. 'Therefore the *Leton* must be *whitened*, and teare the Bookes, least our hearts be broken', writes Artephius; [84] and Maier (p. 236) has an engraving showing the washing of Latona's face with azoth, and the tearing of the books, of which intriguing goings-on the young Apollo and Artemis are interested spectators (Plate 28). 'Latona, therefore, do thou prepare to whiten', says the accompanying epigram; 'and since they do harm, and are of a doubtful sort, with all speed tear the books.' Mylius [85] ascribes almost identical words to Geber.

An English alchemical poem, *The Magistery*, quoted by Ashmole [86] over the initials 'W. B., December 1633', contains two verses which run as follows:

> *Son* and *Moone* in *Hermes* vessell
>   Learne how the *Collours* shew,
> The *nature* of the *Elements*,
>   And how the *Daisies* grow.

> Greate *Python* how *Appollo* slew,
>   *Cadmus* his *hollow-Oake*:
> His *new-rais'd army*, and *Iason* how
>   The *Fiery Steeres* did yoke.

The operations taking place in the Hermetic Vase are here linked in a very suggestive way with the ancient fable of Cadmus and other mythological ideas.

The conjoining of philosophic sulphur and mercury in the Vase of Hermes teaches the operator the nature of the fundamental elements, and shows him how the various colours, or 'flowers', appear in due succession. Apollo represents sophic sulphur; and the python, or serpent, is sophic mercury. The hollow oak to which Cadmus pinned the serpent with his lance, is the Furnace of the

Sages,[87] and the lance represents their Fire. The 'new-rais'd-army' is possibly a metaphoric rendering of 'multiplication'. Jason controlling the fiery bulls is the prototype of the adept manipulating the sacred Fire; similarly, Ben Jonson refers to 'The bulls, our furnace, still breathing fire', in a passage (p. 93) abounding in 'abstract riddles of our stone', based upon classical mythology.

Another poem in Ashmole's collection,[88] by an anonymous author, depicts Daphne and Phoebus as the volatile and fixed (feminine and masculine) principles; since it attempts also to give a complete description of the Work, it merits quoting in full:

A DISCRIPTION OF THE STONE
Though *Daphne* fly from *Phœbus* bright,
    Yet shall they both be one,
And if you understand this right,
    You have our hidden *Stone*.
For *Daphne* she is faire and white:
    But Volatile is she;
Phœbus a fixed God of might,
    And red as blood is he.
Daphne is a Water Nymph,
    And hath of Moysture store,
Which *Phœbus* doth consume with heate,
    And dryes her very sore.
They being dryed into one,
    Of christall flood must drinke,
Till they be brought to a white Stone:
    Which wash with Virgins milke,
So longe untill they flow as wax,
    And no fume you can see,
Then have you all you neede to aske,
    Praise God and thankfull be.

§ 16. THE END OF THE QUEST

The quest of the Philosopher's Stone will stand for all time as the great Odyssey of historical science. The alchemists followed the alluring gleam of this Hermetic *ignis fatuus* through endless philosophic months of forty days; they followed it, as Israel followed the pillar of fire, into alchemical wildernesses, in which they wandered through endless cycles of forty years; faithful to the last, they followed their ' child of fire ' until the full tale of forty generations had at length been told. Then came the end. But the dying spark of the alchemic torch had lit the flame of a new chemistry, which arose phoenix-like from the ashes of the old. The faith of alchemy was the faith that moveth mountains: and who, in this later age, shall say that this faith was held in vain?

' Yet is it fals! ' cries Chaucer's indignant Yeoman,[89] in dismissing his master's claim ' that of a single pound we can make tweye '; and after exposing the knavery of ' this cursed Canoún ', he concludes: ' since God on high wil not that philosóphers signify, how that a man shal come unto this stoon, I counsel for the beste, let it goon '. Nevertheless, in the long-continued quest of that self-same Stone, the sympathetic chemist of to-day discerns the acorn from which has sprung the mighty oak of modern chemistry. ' Does not chemistry promise that instead of seven grains we shall be enabled to raise eight or more on the same soil? ' asks Liebig.[90] ' Is that science not the Philosopher's Stone which changes the ingredients of the crust of the earth into useful products, to be further transformed, by commerce, into gold? Is that knowledge not the Philosopher's Stone which promises to disclose to us the laws of life, and which must finally yield to us the means of curing disease and of pro-longing life? '

So we arrive at a larger vision of the ideal entertained by Paracelsus when he set out, early in the sixteenth century, to widen the scope of alchemy by allying it with medicine; for in modern times the Paracelsian ideal of chemistry in the service of medicine has expanded into chemistry in the service of man. ' If thou strivest to find the Grand Arcanum, ah, cease! ' runs an Italian inscription [91] on an old alchemical engraving (Plate 29). ' Thou wouldest but lose time and trouble.   To a better use doth the Art reserve itself, which is wont to be mistress and leader to the others.'

Leto (Latona), with Apollo and Artemis,
threatened by Python.   From a painted
vase.

CHAPTER IV

THE GOLDEN TRIPOD

This is fairy gold . . . and 'twill prove so.

§ 1. THE HERMETIC MUSEUM

O F the many collections of alchemical tracts, the *Theatrum Chemicum* of Zetzner and the *Bibliotheca Chemica Curiosa* of Manget rank among the largest (p. 116); but one of the most interesting is the smaller *Musaeum Hermeticum*. This was first published by Lucas Jennis at Frankfurt in 1625, as a work of 483 pages, containing nine tracts. A revised and enlarged edition, containing 863 pages, was issued from Frankfurt in 1678; and an English version [1] of this edition appeared in 1893 under the title ' The Hermetic Museum, Restored and Enlarged: Most faithfully instructing all Disciples of the Sopho-Spagyric Art how that Greatest and Truest Medicine of the Philosopher's Stone may be found and held '. The twenty-one tracts embodied in the collection of 1678 bear alluring and characteristic titles, of which the following are representative: ' The Golden Tract concerning the Stone of the Philosophers, by an Anonymous German Adept; The Golden Age Restored . . . by Henry Madathanas; The Sophic Hydrolith, or, Water Stone of the Wise; A Demonstration of Nature, made to the erring Alchemists, and complaining of the Sophists and other False Teachers, Set forth by John de Mehung; [2] A Short Tract, or Philosophical Summary, by Nicholas Flamell; The Glory of the World, or Table

of Paradise; The Book of Lambspring, a noble ancient
Philosopher, concerning the Philosophical Stone, by
Nicholas Barnaud Delphinas; The Golden Tripod; A
Subtle Allegory concerning the Secrets of Alchemy, very
useful to possess and pleasant to read, by Michael Maier;
The Three Treatises of Philalethes; John Frederick
Helvetius' Golden Calf; The All-Wise Doorkeeper, or
a Fourfold Figure '.

These two editions of the *Musaeum Hermeticum* have
a handsome title-page design (Plate 30) bearing the name
of M. Merian, a son-in-law of J. T. de Bry (p. 234).
Four medallions at the sides symbolise the four elements,
air, earth, fire, and water. The top medallion is of
particular interest as a representation of ' the chymic
choir ' (p. 250), and the bottom one is apparently based
upon an emblem published in Michael Maier's *Atalanta
Fugiens* of 1618 (p. 245). Other features of the design
are representations of Sol, Luna, Minerva, Mercury, the
phoenix, and the pelican. Sol and Luna are associated
according to a common convention (Plate 13) with the
zodiacal symbols of the Lion and Crab, and Mercury
bears the caduceus. Minerva (Athena), the goddess of
wisdom, with helmet, golden staff, and shield bearing
an image of Medusa's head, is associated in the engraving with the owl, which was sacred to her in Greek
mythology. The phoenix symbolises renewal, and the
pelican nourishing its young is a figure of revivification
(p. 138).[3]

The dedication of the *Musaeum Hermeticum* of 1625
is followed by a noteworthy drawing (Plate 31), in which
the seven terrestrial metals are symbolised by seven
figures seated in a cave. Above them, on the crust of
the earth, are three other figures carrying the symbols
of fire, water, and ' fiery water '. This last symbol is a six-
pointed star, consisting of two interlacing triangles; it
may be analysed as a combination of the two symbols of
fire and water, or of the four symbols of earth, air, fire,
and water. The design sometimes represents ' universal

matter ', sometimes the *materia prima* of the Stone, and it has also been called the seal of Solomon, or the symbol of wisdom.

The salamander in the top left-hand corner of the drawing denotes fire, and the other three similarly placed drawings stand for air, water, and earth. A Latin verse below the drawing runs as follows: 4 ' The things that are in the realms above are also in the realms beneath; What heaven shews is oft found on earth. Fire and flowing water are contrary one to another; Happy thou, if thou canst unite them: let it suffice thee to know this! ' In agreement with the first statement, which is a clear echo from the Emerald Table of Hermes (p. 54), the seven heavenly bodies depicted in the firmament above influence the production of the seven corresponding metals whose existence is indicated in the ' realms beneath '. The central figure of the ' chymic choir ' sitting in the cave bears a harp, which may be a symbol of the supposed influence of the ' music of the spheres ' upon the process (p. 250). The second idea of uniting fire and water is akin to the conception of ' balancing ' these elements; this is set forth, for example, in a contemporary engraving which depicts Roger Bacon in the act of solving this intriguing problem (Plate 41 (ii); *cf.* p. 16).

The last item in the *Musaeum Hermeticum* of 1678 (not contained in the first edition) is entitled *Janitor Pansophus*, The All-Wise Doorkeeper. It consists of a so-called ' Fourfold Figure ', comprising four large folding plates, the last of which bears the name of Merian: these illustrate, as indicated above (pp. 83-84), the scheme of the Cosmos, the Creation according to the first chapter of Genesis, and the relationship between the Macrocosm and the Microcosm. The fourth plate (Plate 11), which closes the *Musaeum Hermeticum*, ends aptly by quoting that alchemical profession of faith, the Emerald Table of Hermes. These four plates afford an emblematic synopsis of the Hermetic philosophy.

The intelligent student of the early history of chem-

PLATE 39

An Emblem of the Philosopher's Stone.

From a French engraving of the 17th century. Ascribed to Basil
Valentine. (See p. 210.)

PLATE 40

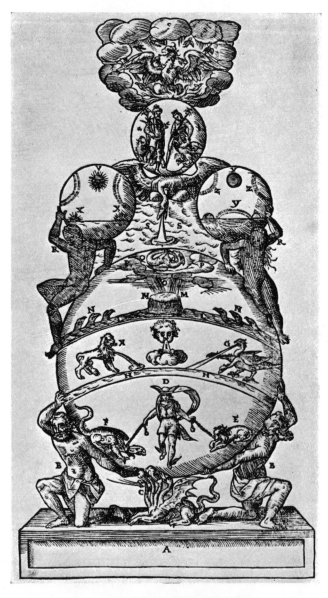

Another sketch of the Philosophic Work.

From *Commentariorum Alchymiae*, Libavius, 1606.

PLATE 41

(i) Hermes the Egyptian.

From *Viridarium Chymicum*, Stolcius, 1624 (reprinted from *Symbola Aureae Mensae Duodecim Nationum*, Maier, 1617). (See p. 223)

(ii) Roger Bacon the Englishman.

From the same. (See p. 223.)

PLATE 42

TRES SCHOLA, TRES COESAR TITVLOS DE.
DIT; HÆC MIHI RESTANT,
POSSE BENE IN CHRISTO VIVERE, POSSE MORI.
MICHAEL MAIERVS COMES IMPERIALIS CON
SISTORII ɛɩc. PHILOSOPH: ET MEDICINARVM
DOCTOR, P. C.C. NOBIL: EXEMPTVS FOR:OLIM
MEDICVS CÆS: ɛɩc:

Count Michael Maier, in 1617.

From *Symbola Aureae Mensae Duodecim Nationum*, Maier, 1617. (See p. 228.)

istry cannot fail to be entertained by the collection preserved in this *Musaeum*. In order to obtain a first-hand
idea of the Hermetic Art, and to assimilate something of
the aims, modes of thought, and methods of expression
of the 'alchemical Artists', one cannot do better than
linger for a while beside one of these masterpieces of a
bygone age. Perhaps the most attractive and illuminating
of all for this purpose is *The Golden Tripod*.

§ 2. THREE NURSLINGS OF THE WEALTHY ART

This work, as the name implies, is a triune composi-
tion, and that peculiarity in itself has an alchemical
significance. The full title is: '*The Golden Tripod*, or,
Three Choice Chemical Tracts, namely: That of Basilius
Valentinus, a Monk of the Benedictine Order; called
Practica, with twelve Keys and an Appendix; The Crede
Mihi, or Ordinal, of Thomas Norton, an English Sage;
The Testament of a certain Cremer, Abbot of West-
minster. Edited by Michael Maier'.

Michael Maier, of whom more will be said later
(p. 228), entrusted the printing of the first edition of the
Tripus Aureus, or 'Golden Tripod', to Lucas Jennis, at
Frankfurt, in 1618.[5] The engraving on the title-page
(Plate 32) shows the Triumvirate in a laboratory, gazing
at a Hermetic Vase which stands upon a Tripod; the
mystic vessel is being heated on a furnace stoked by
Vulcan. An accompanying Latin epigram in *Viridarium
Chymicum* of 1624 (p. 257) hails the 'Big Three' as
'Three Possessors of the Philosopher's Stone', and continues: 'Amiable reader, you behold three nurslings of
the wealthy Art who by their studies have achieved the
Stone. Cremerus in the middle, Norton himself on the
left, Basil, lo, is seen on the right. Pray read their writings and search for the arms of Vulcan [use fire] you who
wish to pluck the apples of the Hesperian ground.'

§ 3. CREMER'S 'TESTAMENT'

These vaunted nurslings may be considered in their chronological order, which may be taken to agree with their sequence in the epigram. The first of them is, in sooth, a very shadowy figure. John Cremer, a reputed brother of the Benedictine Order and Abbot of Westminster, was said to have been initiated into the mystery of preparing the Stone by Raymond Lully, who, according to the tradition, worked with him in Westminster Abbey and in the Tower of London in 1311. Lully's alleged last testament states that while in London he transmuted twenty-two tons of base metal into gold; this was done in order to enable King Edward II (some accounts say Edward III) to prosecute a war against the Turks. This relation is one of many alchemical fables, and, as usual, it contains ingenious touches designed to 'lend an air of verisimilitude to a bald and unconvincing narrative'. Unfortunately, the name Cremer does not occur on the roll of the abbots of Westminster, and Raymond Lully in all probability never visited England; Cremer's *Testament* is pseudonymous, and, moreover, according to modern opinion the alchemical works attributed to Lully are forgeries.

Alchemical writers and editors showed a remarkable laxity in the ascription of authorship to their publications, and their conscience was equally elastic in related matters. Raymond Lully, for example, was credited by some commentators with several thousand treatises! It is significant that the title-page of Cremer's *Testament* is adorned with an engraving (Plate 3) which presumably depicts the author. Illustrations printed from the very same plate appeared in works published by the same firm in the years 1617 and 1624 (pp. 223, 258), with the title 'Thomas Aquinas the Italian'. Similarly, the title-page of Basil Valentine's tract bears a portrait which the editor, Michael Maier, used elsewhere to depict Roger Bacon (Plate 41 (ii)). So much for the reliability of alchemical ascriptions! In introducing Cremer's *Testament*, which is a document of a few pages, the editor warns the reader to be on his guard: 'everywhere a serpent lurks among the flowers. Yet scorn not a friend who spoke as plainly as he might. Beneath the shadowy foliage of words is concealed the golden fruit OF TRUTH.' [6]

Cremer arouses our heartfelt sympathy by his opening statement that he wasted thirty years in reading unintelligible alchemical writings. Then he became acquainted with the 'noble and marvellously learned Master Raymond' in Italy; here, and later during a stay of two years in England, Raymond disclosed to him the whole secret of the Great Mystery. Cremer introduced the Master to King Edward, 'who received him kindly and honourably, and obtained from him a promise of inexhaustible wealth, on condition that he (the King) should in person conduct a Crusade against the Turks, the enemies of God, and that he should thenceforward refrain from making war on other Christian nations'. The King, however, failed to keep his word, and Raymond was forced to flee. According to Ashmole,[7] King '*Edward* the *third* . . . clapt him up in the *Tower*, where he . . . began to study his *Freedom*, and to that end made himselfe a *Leaper*, by which meanes he gained more Liberty, and at length an Advantage of escaping into *France*'.

The *Testament* purports to describe the preparation of 'the living water which constitutes the life of our Art'. Weighed quantities of tartar of good claret, petroleum, 'living sulphur', 'orange-coloured Arsenic', Rabusenum and willow charcoal are heated in a closed vessel for four days. A large number of other materials, including lead and quicksilver, enter into subsequent operations: these, although complicated and prolonged, are clearly described. Here are some extracts from the directions: 'Put it [a sealed glass bottle] in a safe place, and in October you should fill a water-tight box (about one yard in height) with fresh horse-dung, and thrust your glass vessel into it. . . . Shut the lid of the box closely, and never look at the mixture but at the time of the full moon.' At a later stage of the work, a distilling vessel is plunged ' into a wooden box, containing glowing coals of juniper wood [which was reputed to give a lasting fire] and oak, and a twentieth part of iron filings. To test the degree of the fire before inserting the vessel put in it a piece of dry paper. If it catches alight the fire is not too hot, but if the thin shreds which remain of the paper after burning are also consumed, then the heat is excessive, and the door must be opened till the temperature lowers . . . whenever the moon is full . . . remove the cover, and see how the work is progressing. . . . When the mixture turns solid or fixed, its colour should be red of a somewhat dark tinge. . . . It ought to be treated in the manner suggested for forty weeks, beginning on the twenty-fifth of March. By the end of this period the mixture will have become so hard as to burst the vessel. When this happy event takes place, the whole house will be filled with a most wonderful sweet fragrance; then will be the day of the Nativity of this most blessed Preparation.'

Transmutation was accomplished by heating two pounds of lead and two pounds of tin with a quarter of an ounce of the powdered Red Stone resulting from the operations described. When the mixture hardens, after seventy-two hours, ' you have before your eyes the Consummation of the whole work. Mind you lift up your hands in grateful prayer to the Giver of all good gifts. So be it.' Cremer adjures future abbots and priors of his monastery not to betray the secret of the preparation of the Red Dragon's Blood: ' whoever does not observe this my mandate, let his name be blotted out from the Book of Life '.

The concluding sentences of Cremer's *Testament*, which deal with alchemical nomenclature, are not without interest. ' *Rabusenum* is a certain red substance and earth coming forth with water, which flows out of minerals, and is brought to perfection in the month of July in a

glass jar exposed to the heat of the sun for 26 days.'
This may be a reference to chalybeate waters: the alchemists, who often worked under the auspices of the Bull
(p. 211), were attracted by all red materials. ' Magnesia
is the smelted ore of iron.   When the mixture [used in
the Great Work] is still black it is called the Black Raven.
As it turns white, it is named the Virgin's Milk, or the
Bone of the Whale.   In its red stage, it is the Red Lion.
When it is blue, it is called the Blue Lion.   When it is
all colours, the Sages name it Rainbow.   But the number
of such names is legion: and I can only mention these
few.   Moreover, they were only invented for the purpose
of confounding the vulgar, and hiding this mystery from
the simple.   Whenever you meet with a book full of
these strange and outlandish terms and names, throw it
aside at once: it will not teach you anything.'

As evidence of the persistence among English people
at large of the Lully tradition recorded in this tract, it is
of interest to quote an entry which appears in the diary [8]
of Abraham de la Pryme, under date July 11, 1696:
' This day I went to see Madam Anderson, and falling
a talking from one thing to another, shee ran and fetched
me down several old coins to look at, amongst which one
was a rose noble, one of those that Ramund Lully is sayd
to have made [by] chymistry '.   Although the first rose
nobles were struck in 1465, in the reign of Edward IV,
the story was generally accepted.   Ashmole,[9] quoting the
tradition in 1652, refers to an inscription on the reverse
of the noble of Edward III, of 1351 (which was not a
rose noble, as he states): ' *Iesus autem transiens per medium
eorum ibat*, that is, as Jesus passed invisible and in most
secret manner by the midst of *Pharises*, so that *Gold* was
made by *invisible* and *secret Art amidst* the *Ignorant* '.   He
adds, ' *Mayerus* confirmes this '.

Ashmole gilds the refined gold of the tradition by
publishing an engraving [10] alleged to represent a hieroglyphic device which Cremer caused to be painted upon
an arched wall in Westminster Abbey.   Ashmole says:

' I met with it *Limned* in a very *Ancient Manuscript*. . . .
In it is conteyn'd the Grand *Misteries* of the *Philosophers
Stone*, and not more *Popish* or *Superstitious* than *Flamell's
Hierogliphicks* portraid upon an *Arch* in *St. Innocents*
Church-yard in *Paris*; Notwithstanding it has pleased
some, to wash the *Originall* over with a *Plasterer's* whited
Brush.'

§ 4. NORTON'S ' ORDINALL OF ALCHIMY '

Thomas Norton, the author [11] of the middle tract of
the *Golden Tripod*, is said to have been a member of a
Bristol family in the fifteenth century. The chief work
ascribed to him, and the only one which has been printed,
is the *Ordinall of Alchimy*, also known as *Crede-Mihi*, or
' Believe-Me ', which he began to write in 1477. It has
a literary as well as an alchemical interest, since the
original composition took the form of a rhymed English
poem. Curiously enough, it was first published at Frankfurt in 1618 in a Latin [12] translation, as part of Michael
Maier's *Tripus Aureus*. Seven years later a versified
German edition was published by the same firm of Lucas
Jennis, at Frankfurt. The Latin version was reprinted
in Manget's *Bibliotheca Chemica Curiosa* in 1702. Ashmole [13] states that Maier ' *came out of* Germanie, *to live in*
England; *purposely that he might understand our* English
Tongue, *as to Translate* Norton's Ordinall *into* Latin *verse,
which most judiciously and learnedly he did*: *Yet* (*to our
shame be it spoken*) *his* Entertainement *was too course for
so deserving a* Scholler '. He returned to Frankfurt-on-
Main from England in September 1616 (p. 229).

The original English version of the *Ordinall* remained
unpublished until the appearance of Ashmole's *Theatrum
Chemicum Britannicum*, in 1652. This remarkable work
provides a delectable collection of old English alchemical
writings in verse, in the words of the title-page, ' containing Severall Poeticall Pieces of our Famous *English
Philosophers*, who have written the *Hermetique Mysteries*

in their owne Ancient Language '.   Elias Ashmole (1617–1692), the founder of the Ashmolean Museum, Oxford, combined a keen business capacity with the tastes of the scholar, antiquary, and collector.   The *Theatrum* forms the first part of an unfinished collection, and is now rare. The book was a favourite with Sir Isaac Newton, who took a great interest in alchemy.[14]

Ashmole (Plate 33), whose memory should remain green among all English-speaking chemists, points out in his general preface that there was no ' Liberall Science *or any* Feate *concerning* Learning ', in which English writers had not shown ' Arguments *of great* Felicity *and* Wit '.   He felt particularly impelled to record some of their ' Pieces ' dealing with the ' Hermetique Science ', which his diligence had rescued from destruction.   The manuscripts were ' *shrouded in the* Dust *of* Antiquity, *and involv'd in the* obscurity *of* forgotten things, *with their* Leaves *halfe* Worme-eaten.   *And a wonder it is, that (like the* Creatures *in* Noahs Arke) *they were hitherto so safely preserved from that* Universall Deluge, *which (at the* Dissolution of Abbies) *overflowed our greatest* Libraries.'

' *As for the whole* Worke *it selfe,*' he says poetically, ' *it is* sheav'd *up from a few* gleanings *in part of our* English Fields.'   He gives the poetical gleanings the precedence, in the first part of his projected work, since ' *to prefer* Prose *before* Poetry, *is no other, or better, then to let a* Rough-hewen-Clowne, *take the* Wall *of a* Rich-clad-Lady *of* Honour '.

Norton's *Ordinall* is the first and longest of the twenty-nine pieces in Ashmole's collection, in which it occupies pages 1 to 106.   Two other contributions of first-rate importance are ' The *Compound* of Alchymie, A most excellent, learned, and worthy worke, written by Sir *George Ripley*, Chanon of *Bridlington* in *Yorkeshire*, Conteining twelve Gates ', and ' The Tale of the *Chanons Yeoman*. Written by our Ancient and famous English Poet, *Geoffry Chaucer.*'   Ashmole comments upon the injuries done to authors of manuscripts, before the invention of printing, through the negligence of scribes who copied them. The *Ordinall* was printed from ' a very faire ' manuscript (*cf*. Plate 34) ' written in *Velame* and in an auntient *sett Hand*, very exact and exceeding neate. The *Figures* . . . being also most neatly & exquisitely *lym'd*, and better work then that which was *Henry the seaventh's* own *Booke* '; nevertheless, he did not omit to compare it with fourteen other copies.[15]

Thomas Norton shared the secretive instincts of the Hermetic brotherhood, for he concealed his authorship of the *Ordinall* in an acrostical cipher embodied in this work. Ashmole points out that from the first word of the ' Proheme ' (*To*) and certain initial letters of the first six chapters (*Mai*stryefull, *Nor*mandy, *To*nsile, *Of*, *Brise*, *To*wards) of the *Ordinall*, ' we may collect the *Authors* Name and place of Residence: For those *letters*, (together with the *first line* of the seventh *Chapter*) speak thus,·

> ' Toma[i]s Norton of Briseto,
> A parfet Master ye maie him [call] trowe.' [16]

An earlier allusion to this cipher was made by Michael Maier in a work [17] published in 1617.

In a delightfully naïve reference to the fourth capital letter concerned in this cipher, Ashmole remarks: [18] 'The great *Letter* T. set in *pa*. 6. wherein the *Gryphon* is cut, should have been placed the first *Letter* of the *Line*: But this mistake was comitted in my absence from the *Presse*, for which the *Printer* beggs pardon, as also the *Engraver*, for giving the *Gryphons* hinder *Feete*, those *cloven ones* of a *Hogg*, instead of the *ungued pawes* of a *Lyon* '. These ornamented capitals, which were derived by Ashmole from the original manuscript, are reproduced as initial letters in the Proheme (preface) and seven chapters which constitute the present book, and the great T's may be seen on pages xix and 118.

Norton is said to have been initiated (possibly by George Ripley, Canon of Bridlington) into the mysteries of the Red Medicine, within the space of a philosopher's month, when he was but twenty-eight years old.   In the *Ordinall* he writes: [19]

> This *Letter* receiving, I hasted full sore,
> To ride to my *Master* an hundred miles and more;
> And there Forty days continually,
> I learned all the secreats of *Alkimy*.

Norton was one of the earliest of an illustrious line of chemist-poets.   Although he seems to have wooed Calliope less successfully than the Philosopher's Stone, he was usually able to scramble home with a rhyme at the end of his line.   Maier, in a laudatory introduction to readers of the *Golden Tripod*, pays Norton the well-meant but somewhat inelegant compliment of likening the *Ordinall* to the ' fertilising slime ' of the Nile: ' he spreads himself abroad over an immensity of space, that he may fertilise the fields of Alchemy, and rejoice the hearts of its husbandmen '.[20]

In the ' Proheme ' Norton states that the *Ordinall* is a storehouse of secrets for the learned, but he warns the unlearned to ' eschewe the greate deceipts ' of worldly multipliers.   It is clear from Norton's remarks that alchemy was a vogue of the age, and that all classes held a credulous belief in its power to create gold.   Kings and princes, lords ' great of blood ', ecclesiastics from the highest to the humblest—

> And Merchaunts also which dwell in the fiere
> Of brenning Covetise, have thereto desire;
> And *Common workemen* will not be out-laste,
> For as well as *Lords* they love this noble Crafte.

Alchemy, indeed, seems to have appealed to the gambling instinct which is inherent in mankind; mediaeval man patronised the alchemist in much the same spirit as his successor patronises the racehorse and the lottery.

Norton's poem, which is significantly divided into seven chapters, following a Proheme or Introduction, is not free from the common alchemical defects of prolixity and obscurity of expression; but it is not difficult to read,

and it derives an attraction from the archaic language in
which it is written and from the light which it throws upon
the beliefs and practices of fifteenth-century alchemy.   It
can hardly be commended, however, as a practical laboratory handbook for would-be makers of the Philosopher's
Stone.

Norton lays down the principle that the secrets of
' holi Alkimy ' can only be imparted verbally to the chosen
neophyte by a divinely appointed Master—and ' of a
Million, hardly *three* Were ere Ordained for *Alchimy*'.
The recipient must receive the mysteries under the seal of a
' most sacred dreadfull Oath ' to put aside all claims to great
dignity and fame, ' And also that he shall not be so wilde
To teach this seacret to his owne childe '.   The adepts
were unanimous in their determination to maintain esoteric
alchemy as a closed body of knowledge, sacred to the
elect:

> Almighty God
> From great Doctours hath this *Science* forbod,
> And graunted it to few Men of his mercy,
> Such as be faithfull trew and lowly.

A summary, after the manner of Chaucer, of the operations of a certain Tonsile, a misguided ' labourer in the
fire ', discloses some of the materials which he used in
the course of his search for the *ignis fatuus*:

> With weeping Teares he said his heart was fainte,
> For he had spended all his lusty dayes
> In fals Receipts, and in such lewde assayes;
> Of Herbes, Gommes, of Rootes and of Grasse,
> Many kinds by him assayed was,
> As Crowefoote, Celondine and Mizerion,
> Vervaine, Lunara, and Martagon:
> In Antimony, Arsenick, Honey, Wax and Wine,
> In Haire, in Eggs, in Merds, and Urine,
> In Calx vive, Sandifer, and Vitriall,
> In Markasits, Tutits, and every Minerall.

Norton talks of a Stone made from two materials,
male and female in their natures: these are a *White Worke*

or chosen *Markasite* (known in a ruddy form as *Litharge*), and a second Stone called *Magnetia*, 'glittering with perspecuitie, being of wonderfull Diaphanitie'. No other materials are required, save *Sal Armoniak* and the *Sulphur of Mettals*, for preparing the Stone or Elixir. He quotes with approval a description of the Red Tincture attributed to Hermes: 'There lies the snowy wife wedded to her red spouse'. Having disclosed so much, Norton adds somewhat fearfully:

> This secrete was never before this daye
> So trewly discovered, take it for your praye;
> I pray *God* that this turne not me to Charge,
> For I dread sore my penn doeth too large:
> For though much people perceive not this Sentence,
> Yet subtill *Clerks* have too much Evidence;
> For many *Clerks* be so cleere of witt,
> If thei had this ground, thei were sure of it.

Like all alchemists, Norton is interested in colour and colour changes. He resuscitates an older idea by associating dryness and coldness with whiteness (as in quicklime or ice), and heat and moisture with blackness (as in the charring of green wood). These are the extremes of colour, and there are many intermediate tints; but 'Red is last in work of *Alkimy*'. He stresses the importance of maintaining Nature's harmonies of number, weight, and measure. In an accompanying illustration (Plate 34), the adept (apparently Norton himself) is seated at a table upon which stands a balance: divide your material in equal parts, says Norton, 'with subtill balance and not with Eye'. The elements are to be combined arithmetically and also musically (p. 247), because the harmonies of music resemble the proportions of alchemy; they must also be combined under the proper astrological influences.

Norton is at his best when he deals with laboratory operations and administration: this part of his work is much more convincing, and evidently more congenial to him, than his theoretical expositions. He recommends

the hiring of workmen by the day, so that the unsatis-
factory ones may be dismissed quickly; married men
should be avoided, because they dislike this arrangement:
' If I had knowne this, and had done soe, I had avoyded
mickle woe '.   Norton speaks feelingly here, for according to his own statements he suffered much from the
' deceipt ' of servants.   Besides being ill-willed, foolish,
over bold, negligent, ' sleeping by the fire ', and ' filthie
of hands and of sleeves ', some of them were dishonest:

> For when I had my warke well wrought,
> Such stale it away and left me nought.

This was not all, for he continues his tale of woe by relating how a merchant's wife bereft him of the Elixir
of Life—

> To my greate paine and much more woe:
> Soe in this worke there is no more to saine,
> But that every *Joy* is medled with his *paine*.

According to tradition, the fair deceiver was the wife
of William Canynges, who rebuilt the great church of
St. Mary Redcliffe at Bristol—' the fairest, goodliest
and most famous parish church in England ', as Queen
Elizabeth afterwards termed it.   This unauthenticated
story may be an echo from Paris, where Nicolas Flamel
is said to have applied alchemical gold in grandiose building schemes: the idea recurs in a speech of Mammon,
in *The Alchemist*: [21] ' I shall employ it all in pious uses
. . . building hospitals, and now and then, a church '.

Norton insists that laboratory ' ministers ' must be
sober, diligent, discreet, and ' clenly of hands, in Tuching
curious '.   Since the processes must run continuously,
eight servants are usually necessary, divided into two
shifts, so ' That one halfe of them must werke While the
other Sleepeth or goeth to Kerke '.   Moreover, ' all the
Ministers must be Men, Or else thei must be all Weomen '.

The design, material, and size of apparatus must be
regulated in accordance with its uses.   Long vessels are
suitable for some processes, short ones for others; some

are made of lead, some of fireclay, and still others of stone or glass. 'Manie Claies woll leape in Fier, Such for Vessells doe not desire'. Good stone vessels, both impervious to water and resistant to fire, are rarely to be had in England—'such Stones large enough for our intente, were a precious Instrument'. Glass vessels are admirable for retaining 'spirituall matters', and some kinds of glass are harder than others. In stating that the best shape is that which accords most closely with the pattern of Nature, Norton is perhaps referring to the Philosopher's Egg. Here he quotes a saying of Albertus Magnus: 'If God had not given us a vessel, His other gifts would have been of no avail'—he adds 'and that is Glasse'.

Norton displays a particular interest in furnaces. Since so many of the existing furnaces were badly designed, he invented a greatly improved model, in which as many as sixty operations, each requiring a different degree of heat, could be conducted simultaneously:

> The most Commendable Fashion of them all,
> In this Boke portraied finde ye shall.
> One Furnace by me is found of newe,
> Such as Olde Men never knewe.

He invented also another kind of furnace about which he is more reticent:

> This is a new thinge which shall not be
> Set out in Picture for all men to see . . .
> Which suttill Furnace I devised alsoe,
> In which I found manie wonders moe
> Then is convenient at this season to tell.

Norton attributes the highest importance to the control of temperature in the various processes. Would you know the perfect Master?—

> A parfet *Master* ye maie him call trowe,
> Which knoweth his Heates high and lowe.

Another point which Norton emphasises is the great susceptibility of the elements of the Stone, during its preparation, to astrological influences. In his astrological figures (Plate 15) he indicates that the first purification should be undertaken with the Sun in the Archer and the Moon in the Ram, and that the work should be consummated during the conjunction of the Sun and Moon in the Lion. Operations must be protected from malign horoscopic combinations; for the white tincture the moon must be fortunate; the sixth house must be favourable for the servants. 'Trust not to all *Astrologers* . . . for that Arte is as secreat as *Alkimy*'. If you wish to achieve the Great Work, bring your earthly operations into favourable and sympathetic accord with the influences of the heavenly bodies:

> Make all the Premises with other well accord,
> Then shall your merits make you a great Lord.

So we reach the end of the 'greate warke' of this 'parfet Master', Tomais Norton of Briseto; and these are his parting words:

> All that hath pleasure in this Boke to reade,
> Pray for my Soule, and for all both quicke and deade.
> In this yeare of *Christ* One thousand foure Hundred
> seaventy and seaven,
> This Warke was begun, *Honour to God in Heaven.*

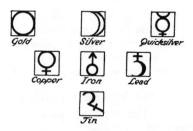

Symbols of the metals, after Glauber.

CHAPTER V

THE MIGHTY KING

'The time has come,' the Walrus said,
'To talk of many things :
Of shoes—and ships—and sealing-wax—
Of cabbages—and kings—
And why the sea is boiling hot—
And whether pigs have wings.'

LEWIS CARROLL

§ 1. THE MYSTERY OF BASILIUS

ASIL VALENTINE, in the true Hermetic tradition, provides one of the most alluring mysteries of alchemy; for although many celebrated alchemical writings were given to the world under this name, Basilius has never been identified.   He is said to have been a member of the Benedictine brotherhood of St. Peter of Erfurt, in the fifteenth century, or earlier still; but the rolls of the Order do not vindicate the statement.   Cremer remains enigmatical; Norton concealed the authorship of his main work in an acrostic; Basil Valentine, the third member of the fellowship of the Golden Tripod, is a 'veiled master'.   Such are the bewilderments of alchemical literature!

The knowledge displayed in some of the works which bear the name of Basilius would establish their author as a remarkable worker, a hundred years or more in advance of his time, if it could be proved that he lived and wrote them in the fifteenth century.   The first printed works [1] ascribed to Basilius appeared towards the opening of the seventeenth century: they were *Von dem grossen Stein der Uhralten* ('Concerning the Great Stone of the Ancients') —Eisleben, 1599, Zerbst, 1602; *Von den natürlichen und übernatürlichen Dingen* ('Concerning Natural and Supernatural Things')—Leipzig, 1603; *De Occulta Philosophia* (or 'The Secret Generation of Planets and Metals')—Leipzig, 1603; and *Triumph Wagen Antimonii* ('The Triumphal Chariot of Antimony')—Leipzig, 1604. These, and other works, passed through various editions in the course of the same century, and enjoyed a very wide circulation; contemporary manuscript versions of some of the writings also exist. Internal evidence shows that some of the works concerned cannot possibly be assigned to the earlier period. The *Triumph Wagen* [2] alludes to 'die neue Franzosen-Kranckheit', or 'Morbus Gallicus', a term which did not reach Germany until about 1500. The same tract mentions the use of antimony in typography, a practice which dates from the closing years of the fifteenth century. A less important tract embodied in *The Last Will and Testament* contains a reference [3] to a new process begun 'in the City of Strasburg . . . in the name of the Holy Trinity, the 19. of *October*, Anno 1605'.

The name Basil Valentine, the 'mighty king' or 'valiant king', has the true alchemical ring, suggestive of the nomenclature of the Philosopher's Stone itself, and it may well have been the pseudonym of an alchemist of the fifteenth century. Possibly even, certain manuscripts of such a writer were utilised by Johann Thölde, the salt manufacturer of Frankenhausen and reputed secretary of the Rosicrucian Order,[4] who 'edited' the *Triumph Wagen* and other works which have been attributed to the supposed Benedictine monk. Many of the ideas and observations are definitely post-Paracelsian, and these have been credited by some critics to Thölde himself, although, as Kopp remarks, it is strange that the discoverer of so many important chemical facts should have published them under another name. It was, however, a well-

PLATE 43

Map of the First English Settlement in Virginia, showing the island of Roanoke and its surroundings. 'The sea coasts of Virginia arre full of Ilands, wher by the entrance into the mayne lãd is hard to finde.'

From the engraving by de Bry in *A briefe and true report of the new found land of Virginia*, by Thomas Hariot (Frankfurt, 1590). (See p. 235.)

PLATE 44

Ægle. Arethusa.   Hespertusa.

ATALANTA
FVGIENS,
Damssouus hoc est,

EMBLEMATA
NOVA

DE SECRETIS NATURÆ
CHYMICA,

Accommodata partim oculis & intellectui, figuris
cupro incisis, adjectisque sententiis, Epigram-
matis & notis, partim auribus & recreationi
animi plus minus 50 Fugis Musicalibus trium
Vocum, quarum duæ ad unam simplicem melo-
diam distichis canendis peraptam, correspon-
deant, non absq; singulari jucunditate videnda,
legenda, meditanda, intelligenda, dijudicanda,
canenda & audienda:

Authore

MICHAELE MAJERO Imperial. Con-
sistorii Comite, Med. D. Eq. ex. &c.

OPPENHEIMII
Ex typographia HIERONYMI GALLERI,
Sumptibus JOH. THEODORI de BRY,

M DC XVIII.

Hercules.

Hippomenes.

Hippomenes.   Atalanta.

Title-page of *Atalanta Fugiens*, 1618.

PLATE 45

'The Woman that washes Clothes.'

From *Secretioris Naturae Secretorum Scrutinium Chymicum*, Maier, 1687 (reprinted from *Atalanta Fugiens*, Maier, 1618).

PLATE 46

A Geometrical Representation of Alchemical Theory.

From *Secretioris Naturae Secretorum Scrutinium Chymicum*, Maier, 1687 (reprinted from *Atalanta Fugiens*, Maier, 1618).

established alchemical custom to ' feign and father ' new writings upon real or imaginary personages with celebrated or high-sounding names, and Thölde may have followed this fashion of securing prestige for his publications. Besides, a writer who calls his chief work ' The Triumphal Chariot of Antimony ' instead of ' Some Contributions to the Chemistry of Antimony and its Compounds ' is only running true to form if he prefer ' The Mighty King ' to plain ' J. Thölde '. Moreover, the legend that the original manuscripts were hidden under a marble slab and disclosed by a supernatural agency (p. 193) is reminiscent of the fabled Emerald Table of Hermes and the devices of the Alexandrian school. There is no doubt also that the compelling name Basil Valentine would attract the attention of the adepts and ensure a wide circle of readers for the books. ' Basilius Valentinus ' was literally a name to conjure with. In the St. Andrews collection there is a copy of the 1624 edition of the *Triumph Wagen Antimonii*, bound in old vellum and stamped in gold with the arms and name of Joachim Enzmulner von und zu Kirchberg, 1636. On the flyleaf, a former owner has scrawled such words and phrases as ' valentibus ', ' silentibus ', ' sal vitalis ', and ' albus intus latens ', in an effort to extract concealed information from the magic cognomen. It is of interest also that the author's name, marked on the back of this volume by its original owner, is given as Thölde(n) and not Basil Valentine. The question of the identity of the author, or authors, of the Basilian writings is important in historical chemistry, and, as Ferguson [5] says, even a partial answer would be of value, ' because since the writings contain apparently first notices of a good many chemical reactions and products, it would be satisfactory to have the date of these settled once for all and assigned to the proper authority '.

Viewing the writings as a whole, Basilius may be described as a semi-scientific practitioner of the gold-making art, with a strong interest in the medicinal action of derivatives of antimony and mercury and other chemical preparations. In some ways he resembles the pseudo-Lully and the mediaeval transmutationists; in others, his writings smack so strongly of Paracelsian doctrines that he and Paracelsus have both been suspected of plagiarism; in still others, he has affinities with the experimental school of Glauber. Yet these three aspects are so closely inter-twined that it is impossible to dissect the writings into the productions of two or three authors. We must regard Basilius as Basilius regards the Philosopher's Stone: ' it is two and three, and yet only one '.

Basilius records many highly important advances in preparative inorganic chemistry, and when he is in the mood he gives excellent working directions for perform-ing the necessary laboratory operations. He is apt, how-ever, to relapse suddenly from the precise and unambigu-ous expression of the chemist into the riotous imagery of the mystical alchemist, often, it seems, from a fear of divulging too freely the secrets of his art.

Among the preparations described by Basilius are metallic antimony, various derivatives of antimony, hydro-chloric acid, solutions of caustic alkali, the acetates of lead and copper, gold fulminate, and many other salts. This work, if original and individual, would stamp Basilius as a practical manipulator and observer of the first rank. Certain admirers of this legendary alchemist have even sought to assign to him some of the credit for the discovery of oxygen, by identifying his ' spirit of mercury ' with the gas, or ' air ', which is expelled from the red calx of mercury (mercuric oxide) by heat. ' It is the Principle to work Metals,' writes Basil [6] of this spirit, ' being made a spiritual Essence, which is a meer Air, and flyeth to and fro without Wings, and is a moving Wind, which after its expulsion out of its habitation by *Vulcan* [fire], is driven into its *Chaos*, into which it entreth again . . . by this Spirit of Mercury all Metals may be, if need requireth, dissolved, opened, and without any corrosive reduced or resolved into their first Matter. This Spirit reneweth both Men and Beast, like the Eagle; consumeth whatsoever is bad, and produceth a great Age to a long life.' A modern chemist might, indeed, descry a dim picture of oxygen in this description; but it is fatally easy to misinterpret an archaic statement in the light of later knowledge.

§ 2. THE TRIUMPHAL CHARIOT OF ANTIMONY

Basilius was particularly interested in the chemistry of antimony, and his most valuable work, *The Triumphal Chariot of Antimony*,[7] has been hailed as the first monograph on a chemical element. Even this ' masterpiece of chemical literature' is not free from suspicion, for as long ago as 1785 it was considered as a possible compilation based upon Alexander von Suchten's *De Secretis Antimonii*, published in 1598, and issued in a later edition by Thölde himself.[8]

Among the apocryphal stories which cluster around this enthusiastic amateur of antimony, there is one, dealing with his favourite metal, of which every chemist will say in his heart, ' se non è vero, è ben trovato '. The legend, which appears to be of French origin, has it that some antimony residues which Basilius had discarded in the course of his numerous experiments were eaten by the monasterial hogs. As a consequence, the animals increased in girth and waxed exceeding fat. Being intensely interested in the medicinal action of chemical preparations, and feeling that a little of the magic medicine might be of equal benefit to his Benedictine brethren, Brother Basilius—carefully excepting himself as a control, in the approved manner of modern science—provided them with a goodly dose. The results were so much in favour of the control that henceforward the metal was known as ' anti-moine ', or ' monk's enemy '.

This joyous story, the veracity of which has been solemnly challenged,[9] probably came into being as a result of a passage in *The Triumphal Chariot of Antimony*:[10]

' Moreover be it known to all, that Antimony doth not onely purge Gold and separate all extraneous additions therefrom, but performs the same operation in the bodyes of men, and other living creatures, which I shall prove by an homely example. If a householder intends to fatten a beast, but especially an hog, let him give him in his meat (three days before he shuts him up) halfe a dragme of crude Antimony, by which means his appetite to his meat will be whetted, and stird up within him, and heel soone grow fat; and if he hath any hurtful quality or disease in his liver, or be leaprous, he shall be healed: This example will seem somewhat grosse to the ears of delicate men; but I intended it for illiterate men, or country people, in whose brains the more subtile Philo-sophy is a meer stranger, that they may discerne that experimentally, which for examples sake I have made use of, that so they may the sooner credit my other writings, wherein I speak more abstrusely: But because theres a great difference between the bodies of men and beasts, I have no intent (by this example here induced) that crude Antimony should be given to men also; because that the beasts are able to bear and concoct much crude meats; which is not permitted to the tender nature and com-plexion of man to doe.'

Much of Basilius' genuine chemical information is contained in *The Triumphal Chariot of Antimony*.   This is preceded by a long-winded discourse in which the writer inveighs against the ' wretched and pitiable medicasters ', in the high Paracelsian vein. ' Fresh wounds can be perfectly cured only by external remedies; but in the case of internal diseases the external application of oils, plasters, ointments, and balms will be of little avail. . . . Come hither, then, ye that claim to be doctors of both branches of Medicine, healers of internal as well as of external diseases: see whether you can make your claim good. . . . For great as is the distance between heaven and earth, so great is the difference between the art of healing internal and external diseases. . . . Good God!

If an examination were held on these points how many doctors of both branches of Medicine would be compelled publicly to declare their ignorance! . . . They do not even trouble to enquire in what way the medicines they prescribe are prepared. Their laboratory, their furnace, their drugs, are at the Apothecary's to whom they rarely or never go. They inscribe upon a sheet of paper, under that magic word " *Recipe* ", the names of certain medicines, whereupon the Apothecary's assistant takes his mortar, and pounds out of the wretched patient whatever health may still be left in him. . . . Antimony, you affirm, is a poison: therefore let every one beware of using it! But this conclusion is not logical, Sir Doctor, Magister, or Baccalaureas; it is not logical, Sir Doctor, however much you may plume yourself on your red cap. . . . Antimony can be so freed of its poison by our Spagyric Art as to become a most salutary Medicine. . . .

'Antimony is one of the seven wonders of the world, and many have written about it without knowing the meaning of their own words; no one before me, and even at the present time no one besides myself, has any real acquaintance with its potency, virtues, powers, operation, and efficiency. If any person could be found, he would be worthy to be drawn about in a triumphal car, like great kings and warriors after mighty and heroic achievements in the battlefield. But I am afraid that not many of our Doctors are in danger of being forcibly placed in such a car ' [11] (Plate 36).

There is much talk of poisons and poisoning in this part of the book, and reference is made to the doctrine of sympathies, or ' like will to like ', which is continually encountered in alchemical literature. So Basilius states: ' Before I attempt to declare the virtue of Antimony, you should know that, although Antimony in its raw state is a deadly poison, yet poison can attract to itself poison more effectively by far than any other heterogeneous substance.

' This assertion is proved by the fact that the body of

an unicorn, which is entirely free from poison, repels every poisonous thing. Place a living spider inside a circle formed by a strip of the skin of an unicorn, and you will observe that the spider will not be able to pass. But if the circle be composed of some envenomed substance, the spider will have no difficulty in crossing the line, which is homogeneous to its own nature.' [12]   Basilius conveniently passes over the difficulty of first catching the unicorn.

The accounts given in the *Triumphal Chariot* of the practical laboratory operations are immediately preceded by the entrance of Antimony, who communicates some essential introductory information: [13] ' It is I, Antimony, that speak to you. In me you find mercury, sulphur, and salt, the three great principles of health. Mercury is in the regulus, sulphur in the red colour, and salt in the black earth which remains. Whoever can separate them, and then re-unite and fix them by art, without the poison, may truly call himself blessed; for he has the Stone, which is called fire, and in the Stone, which can be composed out of Antimony, he has the means of perfect health and temporal subsistence.'

' The signed star ', wrote Paracelsus,[14] probably referring to the regulus of antimony, ' is known to none but the sons of the divine Spagyric Art.' Sharing with him a firm belief in the doctrine of signatures (p. 96), the later alchemists viewed the beautiful crystalline appearance of solidifying antimony with peculiar admiration, reverence, and awe. Indeed, he who first isolated the free element must have experienced a sensation of delight equal to that of Davy, when this young Cornishman of a later age became the first of all humankind to behold a globule of liquid potassium take shape, rise through the crust of potash, and burst into flame.

' Many have esteemed the Signed Star of Antimony very highly,' writes Basilius,[15] ' and spared neither labour nor expense to bring about its preparation. But very few have ever succeeded in realising their wishes. Some have thought that this Star is the true substance of the Philosopher's Stone. But this is a mistaken notion, and those who entertain it stray far afield from the straight and royal road, and torment themselves with breaking rocks on which the eagles and the wild goats have fixed their abode. This Star is not so precious as to contain the Great Stone; but yet there is hidden in it a wonderful medicine, which also may be prepared from it. The Star is compounded in the following way:

'Take two parts of Hungarian Antimony [stibnite, antimony sulphide], and one part of steel; melt with four parts of burnt tartar [potassium carbonate] in an iron basin, such as those in which goldsmiths refine gold. Cool, take out the Regulus [metallic antimony], remove all impurities and scoriæ, pulverise finely, add to it, after ascertaining its weight, three times as much burnt tartar; melt, and pour into [a] basin as before. Repeat a third time, and the Regulus becomes highly refined and brilliant. If you have performed the fusion properly—which is the point of greatest importance—you will have a beautiful star of a brilliant white. The Star is as distinct as if a draughtsman had made it with a pair of compasses.'

This is a remarkably clear and workmanlike account, which would hardly be out of place in a laboratory manual of to-day: the instruction to weigh the reactants is noteworthy. At this point, however, our alchemical Jekyll and Hyde bethinks himself and beats a retreat towards his web of obscurity. 'This Regulus [little king], or Star, may be very often carried through the fire with a stone serpent [heated with a solid corrosive], till at length it consumes itself, and is completely joined to the serpent. The Alchemist has then a hot and ignitable substance, in which wonderful possibilities are latent. It is dissolved into an oil, which should be purified and clarified by transfusion and distillation. Three drops of this oil in two ounces of wine may be administered internally, but with great caution, and not oftener than twice a week.

The proportion should be determined by the peculiar circumstances of the disease, which should, therefore, be known to the physician. This is a remarkable acrid substance, comprehending within itself many arcana, but there is no need to reveal everything at once to the ignorant. Some Arts must be kept secret, in order to stimulate the spirit of enquiry.'

In reading this work, it is noticeable that Basilius loses no opportunity of emphasising the medicinal virtues of his preparations, and this circumstance probably contributed a great deal to the sustained public demand for his works.

The accuracy of the writer's observations and his ability to make shrewd deductions from them are well exemplified in a unique passage from the *Triumphal Chariot*,[16] dealing with some fundamental facts of organic chemistry: ' Wine which has once been changed into vinegar by putrefaction and corruption, can never again produce the spirit of wine, but must always remain vinegar. But when, by means of distillation, the spirit alone is removed from the wine, so that the watery part is separated from the spirit, and the spirit is afterwards sublimed, the wine can never thenceforth become vinegar, even though it were kept a hundred years, but would always remain spirit of wine, just as the vinegar always remains vinegar.

' This change of wine into vinegar is a wonderful thing, for thereby something is actually produced out of the wine which did not before exist in its vegetable essence. In the distillation of wine the first product is a spirit; in the distillation of vinegar the first product is a watery substance, and thus [*i.e.* then] a spirit, as I explained above. Hence the spirit of wine, being itself volatile, renders other things volatile, but the spirit of vinegar fixes and renders solid all medicaments, both mineral and vegetable, so that they attract fixed matter and expel fixed diseases.'

THE MIGHTY KING

§ 3. THE LAST WILL AND TESTAMENT

In harmony with the legend of Basil and the hogs is another which is given prominence in a German work of 1645. An English version was published in London with the following title-page: ' The Last Will and Testament of Basil Valentine, Monke of the Order of St. Bennet, which being alone, He hid under a Table of Marble, behind the High-Altar of the Cathedral Church, in the Imperial City of *Erford*: leaving it there to be found by him, whom Gods Providence should make worthy of it. To which is added Two Treatises. The First declaring his Manual Operations. The Second shewing things Natural and Supernatural. Never before Published in *English*. London, printed by S. G. and B. G. for *Edward Brewster*, and are to be sold at the sign of the *Crane* in St. *Pauls Church-yard*, 1671.' According to the versatile and credulous Olaus Borrichius, who was appointed to the chairs of chemistry and botany at Copenhagen in 1666, the manuscript of the *Last Will* was revealed in the fullness of time through a thunderbolt striking the church. Alchemical ideas were distinguished by a profound conservatism: a similar device—the supernatural opening of a temple pillar—had been used in a work of the pseudo-Democritus, some fifteen hundred years earlier (p. 40), in order to secure the dramatic revelation of a popular alchemical aphorism: ' Nature pleases nature; nature conquers nature; nature produces nature '.[17]

The *Last Will* is a heterogeneous compilation of five books, the last of which is entitled *A Practick Treatise together with the XII. Keys and Appendix of the Great Stone of the Ancient Philosophers*. The first part of this fifth book (pp. 210 to 287) is an English rendering of Basil Valentine's contribution to *The Golden Tripod*. In its entire 534 pages, the *Last Will* probably contains a good deal of material of which Thölde, whose last publication appeared in 1624, had no cognisance. Basil Valentine had become a ' best-seller ' among the alchemical fraternity and the credulous public of the seventeenth century, and successive ' editors ' and publishers took advantage of this fact. Even as early as 1603, Thölde himself complained [18] bitterly that the *Zwölff Schlüssel* (' Twelve Keys '), which he had published at his own expense, had been piratically reprinted. It is hard to realise at the present day that the dust-laden treatises on alchemy which lie neglected in our great libraries once formed the staple literature of a considerable proportion of the reading population; the wide and earnest examination which these books received is evident from the multitudinous annotations which successive owners have left in many of them. Their circulation among ' illiterate men, or country people ' was maintained by the ' certain cures ' for human ills which often abounded in their pages, as well as by their references to gold-making. ' Gold gives power; without health there is no enjoyment ': this saying of Goethe sums up their human appeal.

Many passages recording sound chemical information expressed in clear terms may be found in the *Last Will and Testament*. Here, for example, is an account [19] of the preparation of oil of vitriol, which could only have been written by one who had actually carried out the practical operation:

' If you get such deep graduated and well prepared Mineral, called *Vitriol* [green vitriol, ferrous sulphate], then pray to God for understanding and wisdom for your intention, and after you have calcined it, put it into a well coated Retort, drive it gently at first, then increase the fire, there comes in the form of a white spirit of vitriol in the manner of a horrid fume, or wind, and cometh into the Receiver as long as it hath any material in it . . . if you separate and free this expelled spirit well and purely *per modum distillationis*, from its earthly humidity, then in the bottom of the glass you will find the treasure, and fundamentals of all the Philosophers, and yet known to few, which is a *red Oyl*, as ponderous in weight, as ever

any Lead, or Gold may be, as thick as bloud, of a burning
fiery quality.'

At this point Basilius passes from plain chemical to
mystical alchemical language. He goes on to describe
the oil as 'that true fluid Gold of Philosophers, which
nature drove together from the three principles, wherein
is found a spirit, soul, and body, and is that *philosophick
Gold* . . . [which] receiveth its first birth and beginning
from a heavenly water, which in due time is poured down
upon the earth. In these together driven goldish waters
lieth hid that true bird and *Eagle*, the King with his
heavenly *Splendor* together with its clarified Salt.' This
statement of the importance of the vitriol of heavenly
origin, and its derived oil, finds a symbolical expression
in the celebrated 'vitriol acrostic', dealing with the preparation of the Philosopher's Stone (Plate 18 (ii)).[20]

A description [21] of the preparation of lead acetate,
or sugar of lead, contains an allusion to the famous
'Balneum Mariæ', or 'Bain-Marie'.[22] 'Note this for
a *memorandum*, if distill'd pure Vinegar be poured upon
destroyed *Saturn*, and is kept warm in *Marie's-bath*, it
loseth its acidity altogether, is as sweet as any Sugar, then
abstract two, or three parts of that Vinegar, set it in a
Cellar, then you will find white transparent stones, like
unto Crystals, these are an excellent cooler and healer of
all adust and inflamed Symptoms.'

Finally, here is a method [23] for preparing a solution
of caustic potash, described by one who had obviously
experienced the action of this 'spirit' upon the skin :
'Quick-lime is strengthned, and made more fiery, and
hot, by a pure and unsophisticated spirit of Wine, which
is often poured on it, and abstracted again, then the white
Salt of Tartar [potassium carbonate] must be grinded
with it, together with its additionals, which must be dead,
and contain nothing, then you will draw a very hellish
spirit, in which great mysteries lye hid. How this spirit
is gotten, I told it, observe it, keep it, take it for a fare-
well.'

PRELUDE TO CHEMISTRY

§ 4. The Twelve Keys

Maier's selection of the *Twelve Keys* as the first and longest item in the *Golden Tripod* shows that he attached more importance to the esoteric writings of Basilius than to his more exoteric contributions to alchemy. This valuation accorded with the general estimate of the times; for this tract, with its extravagant language and bizarre illustrations, appears to have been the most popular of all the works attributed to Basilius. It is not surprising, however, that a work so rich in allegory, leavened with mystical religion, should make a strong appeal to an age which produced John Bunyan.

Maier's Latin translation was made from the German edition published at Frankfurt in 1611. There was an earlier German edition,[1] and numerous later issues were made in various languages.[24] The twelve emblems also have appeared in many variants. An English translation, provided with very crude drawings of the emblems, was included in the *Last Will and Testament*, with the following title-page: 'A Practick Treatise Together with the XII. Keys and Appendix of the Great Stone of the Ancient Philosophers. Written and left by *Basilius Valentinus* a *German* Monke of the Order of St. Bennet. London, Printed by S. G. & B. G. for *Edward Brewster*, at the *Crane* in Saint *Pauls Church-yard*, 1670.'

Maier's opinion of this work is given in an introductory epigram contained in the *Golden Tripod*. In flowery and allusive language he states that in this book Basilius has disseminated greater wealth than all the riches of the Inds; he has endowed the fair fields of Germany with golden fruit plucked in the Hesperian garden, and with the golden fleece borne away from Colchis by mighty toil. 'Here is something for you to admire and imitate', says Maier,[25] adding cryptically, 'Only seek it at the bottom of the vessel, or you will wander astray'.

In his preamble to the *Twelve Keys*, Basilius warns

the seeker after the Stone that it cannot be derived from materials unable to resist fire. Although it is capable of growth, it is not of vegetable or animal origin. It arises from the seed of metals, formed in the following way:[26]

'A Celestial influence . . . descendeth from above, and mixeth itself with the Astral properties . . . these two beget an earthly substance, as a third thing, which is the beginning of our seed.' This is the interpretation of the hard saying that the Stone ' proceedeth and ariseth from two and from one thing [male and female, two in one], which containeth a third [the seed of metals] concealed '. From these three things arise the ' elements '—water, earth, and air. These ' elements ', under the influence of a subterranean fire, bring forth the three first Principles —' an intrinsick Soul, an impalpable Spirit, and a corporeal and visible Essence '. In time, under the continued action of fire, the three first Principles are changed into palpable Mercury, Sulphur, and Salt, from which three ' by commixtion . . . there is made a perfect body, as Nature would have it, and its seed is chosen and ordained by the Creator. . . . If there be a Metalick Soul, a Metalick Spirit, and Metalick form of Body, there must also be a Metalick Mercury, a Metalick Sulphur, and a Metalick Salt, which of necessity can produce no other than a perfect Metaline Body. . . . You need not look for our metallic seed among the elements. It need not be sought so far back. If you can only rectify the Mercury, Sulphur, and Salt (understand, those of the Sages) until the metallic spirit and body are inseparably joined together by means of the metallic soul, you thereby firmly rivet the chain of love, and prepare the palace for the coronation.' [27]

' If you do not understand this that you ought to understand,' adds Basilius, ' you are not adepted for Philosophy, or God concealeth it from thee.' A little later, however, he relents somewhat and remarks that if the theoretical part of his treatise has proved too difficult, perhaps a practical account of his achievement of the

Stone of the Ancients will be easier to follow. 'Consider it well, and with diligent and frequent reiteration throughly read my XII Keys, and so proceed, as I shall here teach and instruct you, fundamentally by way of Parable.'

Straightway, mounting one of the most mettlesome of his mystical hobby-horses, Basilius clutches the reins, takes a firm seat, and rides away to attend an Alchemical Garden Party beside which the Mad Tea-Party of Alice pales into platitudinous insignificance. Here he encounters [28] '*Mercury*, so proud that he scarce knew himself; he cast off his Eagles wings, and himself swallowed up the slippery tail of the Dragon, and offered battel to *Mars*. Then *Mars* gathered his Champions together, and gave command that Mercury should be imprisoned, to whom *Vulcan* was appointed Gaolor, until he should be freed by some of the feminine kind.' In a fierce speech, delivered at a hastily convened council of the planets (metals), Saturn then began to assail 'inconstant Mercury', to whose carelessness and negligence he attributed his own infirm and corruptible body, his black colour, and his uncomfortable habits. 'Therefore, my Lords, I pray you, revenge my quarrel on him, and seeing that he is already in Prison, kill him, and let him putrefie there, until not one drop of his bloud be any more found.' Saturn was ably seconded by 'brown *Jupiter*', and then came Mars to Vulcan the gaoler, with his naked Sword of many colours, to play the part of Lord High Executioner. Unluckily for Mercury, his friends of the feminine kind arrived late. They were 'Luna . . . a beautiful and white shinning Woman, in a long Robe of a silver colour . . .' and 'Dame *Venus* in a garment of pure red, interwoven with green, of a most beautiful countenance, a most graceful and pleasant speech [in the Chaldean language], and most amiable gesture, bearing most fragrant flowers in her hand'.

Having effectively 'fixed' Mercury, the Lord High Executioner began to pay unwelcome professional attentions to Sol, greatly to the dismay of the 'beautiful and shinning' Luna, who was by this time in tears. Fortunately for Sol, the proceedings were interrupted at this critical point by the sudden and unannounced intrusion of ' a great Animal with many thousands of young ones, driving away and expelling the Executioner' (Plate 56 (ii)). The Animal and his progeny all clamoured for food, ' and they drank of the former incombustible Oyl, and they did easily digest their meat and their drink, and they had many more young ones than before, and this happened often '. The scene had become uncomfortably congested when ' at last came forth a certain Old Man, his Beard and Hair as white as Snow '. From this description, the distinguished visitor was presumably the mysterious Senior or Elder (pp. 268, 276), a favourite alchemical *deus ex machina*. ' This man ascended the Chair, and exhorted the Assembly there met to be silent.' He then delivered a long speech, in the course of which he surveyed a comprehensive range of subjects extending from alchemical cabbages to alchemical kings; he drew attention also to the striking fact that ' the Phoenix of the South hath snatcht away the heart out of the breast of the huge beast of the East, as hath the bird of the South, that they may be equal; for the beast of the East must be bereaved of his Lions skin, and his wings must vanish, and then must they both enter the Salt Ocean, and return again with beauty '. After making a few other judiciously selected remarks, including a topical reference to the Emerald Table of Hermes, the self-constituted Chairman considerately ' vanished away before their eyes '. This concluded the business of the meeting; for ' this Speech being ended ', without any reference to the serious housing problem which had been so suddenly created, ' every one returned unto his own home '. It was high time; for—as the reader is left to infer—

> 'Twas brillig, and the slithy toves
>   Did gyre and gimble in the wabe;
> All mimsy were the borogoves,
>   And the mome raths outgrabe.[29]

Incidentally, alchemical literature abounds in illustrations of slithy toves and mimsy borogoves: some of them are shown in Plate 5.

With this illuminating introduction, the tract [30] continues: ' Now follow the XII Keys of BASILIUS VALENTINUS, wherewith the Doors are opened to the most Ancient Stone of our Ancestors, and the most secret Fountain of all Health is discovered '.

Each of the Twelve Keys consists of a detailed allegorical description of an accompanying emblem. In order to illustrate the nature of the Keys as clearly as possible, the extracts quoted below have been taken from a modern English translation [31] of the original Latin text of the *Tripus Aureus*; in each case the extract forms only a fraction of the complete description. A contemporary précis of certain Keys has been added in the form of an English translation of the complete Latin epigram which was printed with the corresponding emblem in the *Viridarium Chymicum* of 1624 (p. 256).[32]

The First Key of Basilius (Plate 37 (i))

' Let the diadem of the King be of pure gold, and let the Queen that is united to him in wedlock be chaste and immaculate.

' If you would operate by means of our bodies, take a fierce grey wolf, which, though on account of its name it be subject to the sway of warlike Mars, is by birth the offspring of ancient Saturn, and is found in the valleys and mountains of the world, where he roams about savage with hunger. Cast to him the body of the King, and when he has devoured it, burn him entirely to ashes in a great fire. By this process the King will be liberated; and when it has been performed thrice the Lion has overcome the wolf, who will find nothing more to devour in him. Thus our Body has been rendered fit for the first stage of our work.'

It is impossible to give precise and unequivocal interpretations of the emblems, if only on account of the vary-

PLATE 47

Death and Resurrection of the King.

From *Secretioris Naturae Secretorum Scrutinium Chymicum*, Maier, 1687 (reprinted from *Atalanta Fugiens*, Maier, 1618). (See p. 241.)

PLATE 48

EMBLEMA XIX. *De secretis Naturæ.*

Si de quattuor unum occidas, subitò mortuus omnis erit.

EPIGRAMMA XIX.

Bis duo stant fratres longo ordine, pondera terra
   Quorum unus dextra sustinet, alter aquæ :
Aëris atque ignis reliquis est portio, si vis
   Ut pereant, unum tu modo morte premas:
Et consanguineo tollentur funere cuncti,
   Naturæ quia eos mutua vincla ligant.

POETÆ

The Four Brothers.

From *Secretioris Naturae Secretorum Scrutinium Chymicum*, Maier, 1687 (reprinted from *Atalanta Fugiens*, Maier, 1618)

PLATE 49

Oedipus and the Sphinx.

From *Secretioris Naturae Secretorum Scrutinium Chymicum*, Maier, 1687 (reprinted from *Atalanta Fugiens*, Maier, 1618).

PLATE 50

The Stone of Saturn.

From *Atalanta Fugiens*, Maier, 1618.

ing meanings of alchemical signs and figures; for although the alchemists clung fervently to the gold standard, their stock of symbols formed a fluctuating and unstable currency. As we have seen (p. 102), the King, or Sol, may represent either esoteric ' sophic sulphur ' or exoteric gold; and the Queen, or Luna, is ' sophic mercury ' in esoteric alchemy and silver in exoteric alchemy. The King is clad in red and the Queen in white, or blue, and their alternative symbols are often a red rose and a white lily.

The alchemical wolf in general represents a corrosive or ' biting ' agent, sometimes an acid. In this case, the grey wolf is clearly antimony, which was known to the alchemists as *lupus metallorum*, or ' wolf of the metals ', because it ' devoured ', or united with, all the known metals except gold. On account of its use in purifying molten gold—the impurities being removed in the form of a scum—antimony was also called *balneum regis*, the ' bath of the King '. Moreover, its ' appetite ' for metals was likened to the mythological appetite of Saturn for infants (p. 243): antimony was therefore sometimes called ' the sacred lead ', or ' lead of the philosophers '.[33]

Although there is no mythological authority for showing Saturn with a wooden leg, the Ancient of Days in the picture is Saturn, or lead. Saturn takes over the mythology of Kronos, who is often represented as an old man holding a curved implement shaped like a scythe or sickle. The wooden leg probably symbolises the slow movement of Saturn through the sky. Vulcan, the fire-god, was shown with a wooden leg for another reason (p. 143).

The following ' interpretation ' is adopted for the first emblem of Basilius in the *Viridarium*: ' From the yellow metal let a crown be made; let the modest bride be united to her Bridegroom. Then hand over the King to be eaten by the ravening Wolf, and that three times; and stoutly burn the Wolf in fire. Hence the King will come forth cleansed from every spot, who can renew thee with his own blood.'

There is little doubt that the emblem, which closely resembles that of Plate 56 (i), represents the preparation of the 'primitive materials' of the Philosopher's Stone (p. 131). The liberated or revivified King symbolises purified gold. This yields in turn sophic sulphur, or the seed of gold, which is then used in preparing the Philosopher's Stone by uniting it with sophic mercury, or the quintessence of the metals, derived from purified silver (the immaculate Queen). Coming to practical details, the gold is purified by three successive fusions with antimony, in the crucible over which the wolf is leaping; and the silver is 'fixed' (p. 63) by heating it with lead, in the cupel shown in front of the Queen and beside Saturn. The scythe is here a symbol of fire, used in the cupellation. The three flowers held by the Queen symbolise the triple purification of the King, and her fan of peacock's feathers is allusive to the appearance of colours in the later processes of the Great Work (p. 145). The emblem is thus a pictorial expression of the application of the sulphur-mercury theory in the earlier stages of the preparation of the Philosopher's Stone.

The Eighth Key of Basilius (Plate 37 (ii))

Basilius used the common alchemical allegory of death, followed by resurrection, to represent steps in the preparation of the Stone and the ennobling of the baser metals (p. 138). The emblems of his fourth and eighth Keys depict graveyard scenes. He states that ' neither human nor animal bodies can be multiplied or propagated without decomposition; the grain and all vegetable seed, when cast into the ground, must decay before it can spring up again; moreover, putrefaction imparts life to many worms and other animalculae . . . if bread is placed in honey, and suffered to decay, ants are generated. . . . maggots are also developed by the decay of nuts, apples, and pears.

' The same thing may be observed in regard to vegetable life. Nettles and other weeds spring up where no such seed has ever been sown. This occurs only by putrefaction. The reason is that . . . the seed is spiritually produced in the earth, and putrefies in the earth. . . . Thus the stars and elements may generate new spiritual, and, ultimately, new vegetable seed, by means of putrefaction. . . .

' That there can be no perfect generation or resuscitation without the co-operation of the four elements, you may see from the fact that when Adam had been formed by the Creator out of earth, there was no life in him, until God breathed into him a living spirit. Then the earth was quickened into motion. In the earth was salt, that is, the Body; the air that was breathed into it was mercury, or the Spirit, and this air imparted to him a genuine and temperate heat, which was sulphur, or fire. Then Adam moved, and by his power of motion shewed that there had been infused into him a life-giving spirit. . . . Water was incorporated with the earth. Thus living man is an harmonious mixture of the four elements.'

The emblem indicates that the material of the Great Work dies, putrefies, or blackens, in the Philosopher's Egg, but later undergoes revivification, albification, or whitening. The crows denote putrefaction; and the angel sounding the last trump is symbolic of the resurrection of man, thus also of the revivification of metals. The two crossbowmen aiming at the target possibly represent the adept and the puffer: the first scores a bull, but the second wanders in the outers or misses his aim.

The Ninth Key of Basilius (Plate 38 (i))

The emblem of the ninth Key is perhaps the most bizarre of all the pictorial delineations of Basil Valentine. ' Saturn, who is called the greatest of the planets, is the least useful in our Magistery. Nevertheless, it is the chief Key of the whole Art, howbeit set in the lowest and meanest place. Although . . . it has risen to the loftiest height . . . its feathers must be clipped, and itself brought down to the lowest place, from whence it may once more be raised by putrefaction, and the quickening caused by putrefaction, by which the black is changed to white, and the white to red, until the glorious colour of the triumphant King has been attained. . . . This transmutation is begun, continued, and completed with Mercury, sulphur, and salt.'

The ninth Key thus deals with lead, the Cinderella of the alchemists. The Humpty-Dumpty of the curious drawing is Saturn having his great fall, and simultaneously losing his feathers or volatility, as the black crow attains domination over the eagle. He is being replaced by the figure of a more worthy metal, probably Luna (sophic mercury), who carries the ascensive sign of the white swan and bases her feet upon the phoenix, a symbol of revivification and of the Stone. The outside square represents, as usual, the four elements; the double circle is the sign of the twofold mercurial substance; and the three serpents emerging from hearts are the symbols of the *tria prima*—mercury, sulphur, and salt.

The emblem apparently signifies that argentiferous lead upon cupellation becomes fixed, yielding a residual bead of silver which is not affected by further heating. Putrefaction here denotes the ' killing ' of the lead by oxidation in the cupel. The sophic mercury obtained in turn (p. 132) from the residual silver is then used, as shown in the first Key, in preparing the Philosopher's Stone, that is, ' the triumphant King '.

The description of this emblem is remarkable for its wealth of imagery. There are references to Charity, Astronomy, Rhetoric, Hope, Geometry, Courage, Arithmetic, Temperance, Grammar, and Dialectic. The closing pronouncement is to the effect that ' the present state of things is passing away, and a new world is about to be created, and one Planet is devouring another spiritually until only the strongest survive. Let me tell you allegorically that you must put into the heavenly Balance the

Ram, Bull, Cancer, Scorpion, and Goat.   In the other
scale of the Balance you must place the Twins, the
Archer, the Water-bearer, and the Virgin.   Then let
the Lion jump into the Virgin's lap, which will cause the
other scale to kick the beam.   Thereupon, let the signs
of the Zodiac enter into opposition to the Pleiads, and
when all the colours of the world have shown themselves,
let there be a conjunction and union between the greatest
and the smallest, and the smallest and the greatest.'

This is Ercles' vein, and—although Pernety [34] has
attempted an interpretation—the reader cannot but feel
grateful to Basilius for a considerate observation which
he has placed in the middle of this Key: ' This will seem
unintelligible to many, and it certainly does make an
extraordinary demand upon the mental faculties '.   There
is method in this madness, however: the obscurity of
expression is imperative, ' because the substance is within
the reach of everyone, and there is no other way of keeping
up the divinely ordained difference between rich and
poor '.   If alchemy had survived, it would have become
political.

The Twelfth Key of Basilius (Plate 38 (ii))

The emblem [35] of the last Key of Basil Valentine has
been reproduced in many works, in the seventeenth and
eighteenth centuries and also in modern times.   It has
been interpreted in a great variety of ways.   Basilius himself, assuming that the would-be adept has now ' achieved
the Stone ', introduces the emblem by remarking that
' he that possesses this tincture, by the grace of Almighty
God, and is unacquainted with its uses, might as well
not have it at all.   Therefore this twelfth and last Key
must serve to open up to you the uses of this Stone.   In
dealing with this part of the subject I will drop my para-
bolic and figurative style, and plainly set forth all that is
to be known.

' When the Medicine and Stone of all the Sages has

been perfectly prepared out of the true virgin's milk [p. 157], take one part of it to three parts of the best gold purged and refined with antimony, the gold being previously beaten into plates of the greatest possible thinness. Put the whole in a smelting pot, and subject it to the action of a gentle fire for twelve hours; then let it be melted for three days and three nights more.

'For without the ferment of gold no one can compose the Stone or develop the tinging virtue. For the same is very subtle and penetrating if it be fermented and joined with a ferment like unto itself; then the prepared tincture has the power of entering into other bodies, and operating therein. Take then one part of the prepared ferment for the tinging of a thousand parts of molten metal, and then you will learn in all faith and truth that it shall be changed into the only good and fixed gold. For one body takes possession of the other; even if it be unlike to it, nevertheless, through the strength and potency added to it, it is compelled to be assimilated to the same, since like derives origin from like. . . .

'*O Beginning of the first Beginning, consider the end.*
O End of the last End, see to the Beginning.'

The open allusion to antimony in this Key serves as a connecting link between Basilius of the *Chariot* and Basilius of the *Keys*. The Twelfth Key indicates that the Stone must be incorporated with gold before it can be used in projection for the production of new gold. Just as the lion of the emblem changes the serpent into its own substance by devouring it, so the Stone or powder of projection—often known as the Red Lion—changes imperfect metals into its own substance, that is, gold. 'If the high-souled Lion chances to devour the Snake', says the epigram in the *Viridarium Chymicum*, 'then Mercury will give you a thousand flowers [the colours of the Great Work]. For without a ferment the Stone worketh not to make gold; but joined thereto, at its entering in, it can tinge many things. If you use this agent, you will discern hidden things, and see God favourable to your prayers.'

From the above comments on four of the Keys, it is seen that they conform to their author's claim to open the Doors to ' the most ancient Stone of our Ancestors '. They deal emblematically and allegorically with the operations of the Great Work, although no logical sequence is evident in their arrangement.

An Appendix [36] following the Twelfth Key is intro-

Fig. 13.—A Synthetic Emblem of the Great Work.
From *Viridarium Chymicum*, Stolcius, 1624 (reprinted from *Tripus Aureus*, Maier, 1618).

duced by an emblem (Fig. 13) which must be regarded as an alchemical *multum in parvo*, since it contains a compressed synthetic expression of the Great Work: it summarises the processes which are darkly indicated in the Keys as necessary for the preparation of the Philosopher's Stone. It is therefore strange that a work [37] on historical chemistry should dismiss so significant an emblem as a mere representation of the process of distillation; but unfortunately many similar examples might be quoted of the undiscerning reproduction of alchemical emblems in modern publications. The emblem here concerned has nothing to do with distillation. The enclosing square represents the four elements; the triangle and the three crowned serpents stand for the *tria prima* — sophic sulphur, mercury, and salt. The two intersecting circles suggest the dual influence which Basilius defines earlier (p. 197) as ' celestial influence ' and ' astral properties ', but which is often more simply regarded as a conjunction of the masculine and feminine principles; the same idea is conveyed by the hermaphroditic ' Rebis ', or ' Two-Thing ' (p. 134). One of these circles encloses the dragon's paws and the other his wings, thereby signifying, it may be, the fixed and volatile, or even the elements earth and air. The dragon is the proximate material of the Stone (p. 131); the large circle symbolises universal matter; when attached to a stem, or tube, as in this emblem, it delineates the Philosopher's Egg.

' I have already indicated ', explains Basilius, ' that all things are constituted of three essences—namely, mercury, sulphur, and salt—and herein I have taught what is true. But know that the Stone is composed out of one, two, three, four, and five. Out of five—that is, the quintessence of its own substance [the Aristotelian *quinta essentia*, or ' fifth essence ']. Out of four, by which we must understand the four elements. Out of three, and these are the three principles of all things. Out of two, for the mercurial substance is twofold. Out of one, and this is the first essence of everything which emanated from the primal fiat of creation.'

An interesting expression of the same ideas occurs at the beginning of a late edition of the *Novum Lumen Chemicum* (' New Light of Alchemy '), printed at Nuremberg in 1766 (Fig. 14). The four elements, fire, air, water, and earth, are shown as proceeding directly from God; the three principles, sulphur, salt, and mercury, from Nature; the two ' seeds ', the male and female—Sol and Luna, or sulphur and mercury—from the metals; and the one Tincture, from art. The appended verse relates how the four elements give rise to the three ' beginnings ', which in turn produce the masculine and feminine species: from these—the Sun and Moon—is born the royal son, whose like is not to be found in the whole world.

vier Elementen.	Drey Anfänge.	zwen Saamen.	eine Frucht.
Feuer. △ Luft. △ Waſſer. ▽ Erde. ▽	Schwefel. ♀ Salz. ♁ Mercur. ☿	Männlein. ☉ Weiblein. ☽	Tinctur. ☿
von Gott.	der Natur.	der Metallen.	der Kunſt.

Wer dieſe Tafel recht verſtehet,

Sieht wie eins aus dem andern gehet,

Erſtlich ſteckt alles in vierdter Zahl,

Der Elementen überal.

Daraus die drey Anfänge entſpringen.

Welche zwey Geſchlechter herfür bringen,

Männlich, Weiblich, von Sonn und Mon,

Daraus wachſet der kayſerliche Sohn,

Dem auf der Welt gar nichts iſt gleich,

Und übertrift all Königreich.

FIG. 14.—An Arithmetical conception of the Stone. From *Michaelis Sendivogii, Novum Lumen Chemicum*, Nuremberg, 1766.

These examples of simple arithmetical and geometrical conceptions are typical of the influence of the Pythagorean emphasis on number, which is as apparent in alchemical doctrine as in the thought of Copernicus and Kepler. This interesting aspect of alchemy finds many reflections in folklore, as well as in widespread superstitions regarding the magic qualities of certain numbers and geometrical shapes. A faint alchemical echo may even be detected by an imaginative ear in the widespread 'Dilly Song' and its many variants: [38]

Seven are the seven stars in the sky,[a]
And six are the six broad waiters.[b]
Five are the flamboys[c] under the boat,
And four are the gospel makers.
Three of them are thrivers,[d]
And two and two are lily-white babes [e] a-clothed all in green, O!
One and One is all alone, and evermore shall be so.

The esoteric atmosphere of the *Twelve Keys* is maintained in other works ascribed to Basil Valentine, and some of these abound in quaint symbolic illustrations, which, like those of the *Keys*, were received with superstitious credence in the seventeenth century. One of the most interesting of such works is a short, illustrated tract published in several forms [39] from 1613 onwards, and reissued in Paris, in 1659, under the title, *Azoth, ou le moyen de faire l'Or caché des Philosophes*; this is considered below (p. 268).

This short review of some of the leading works ascribed to Basilius Valentinus may end appropriately with a reference to two emblems, sometimes associated with his name, which afford synopses of the processes leading to the Stone. The first of these occurs in *Azoth*, and is described elsewhere (p. 105). The second, which is reproduced below (Plate 39) from a seventeenth-century French engraving, shows the cubic Stone in the midst of seven flowers representing the heavenly bodies and metals. Above it is the mercurial sign, flanked by Sol and Luna. Then comes sulphur, enclosing the Phoenix, and surmounted by the crowned King marked with the signs of gold, silver, and mercury, and holding aloft the compasses

---

[a] Sun, moon, and five planets.
[b] Water-pots used in the transmuting miracle at Cana.
[c] Waves.   [d] Wisers, or Magi.
[e] The Gemini, or Twins.

and scythe; on his left is a representation of fire, and the symbol on his right (so suggestive of the centric formula for benzene!) probably signifies the blood of the Pelican or Red Dragon, mentioned in the third Key. An elaborated form of this engraving, appearing in *Le Triomphe Hermétique*,[40] published at Amsterdam in 1699, bears the inscription: *De cavernis metallorum occultus est, qui Lapis est venerabilis. Hermes.* (' From the caverns of the mines it comes, the hidden one, which is the venerable Stone '). In it the cubic Stone is replaced by the Hermetic Vase; the Hermetic Stream of ' heavy water ' and the element Fire issue from subterranean caves; and a decorative zodiacal design in the firmament indicates that the Work must be begun when the Sun is in the sign of the Ram, and that its consummation must be reached in the Bull, ' when the fortieth dawn returns ' (p. 271).

Another symbol of the Stone, showing Sol, Luna, the Lion, Eagle, and Salamander. From *Viridarium Chymicum*, Stolcius, 1624.

CHAPTER VI

A MUSICAL ALCHEMIST

He does not appear to have been included among the adepts, and he is now almost forgotten. His chemical knowledge is buried in a multitude of symbols and insoluble enigmas, and believers in spiritual chemistry will not derive much comfort or profit from his writings.

A. E. WAITE

§ 1. THE FRANKFURT EMBLEMS

HE agreeably piquant illustrations, so racy of the alchemical soil of the early seventeenth century, are perhaps the most spectacular of many features of interest in the works attributed to Basil Valentine. The Twelve Keys of Basilius, in particular, were handled repeatedly by new artists and interpreters, who provided them with an honoured place in other publications of this time. Judging from the popularity of Ben Jonson's play, *The Alchemist*, which was first produced in 1610, there was a considerable public throughout the seventeenth century for attractive expositions of alchemy: frequenters of the playhouses of those days were familiar with that technical language of contemporary alchemy and astrology which is so much ' heathen Greek' to the modern playgoer. Thus, the production of titillating alchemical works, abounding in pictorial illustrations, came as a response to a wide demand.

The first issue of the Basilian emblems appears to have been made from Frankfurt-on-Main (p. 196); and other notable alchemical works with similar pictorial

embellishments were published, either in their original
or later editions, from this famed centre of early book
illustrations.   We may therefore conveniently group this
outstanding series of alchemical emblems under the name
of ' the Frankfurt Emblems '.   These intriguing enigmas
—so suggestive of the later ' hieroglyphs ' of Old Moore
and Zadkiel—probably reached their zenith in certain
works published in the second decade of the seventeenth
century by that Michael Maier who has already been
mentioned (p. 169) as the editor of the *Golden Tripod*.

§ 2. LIBAVIUS ON ENIGMAS OF THE STONE

Precursors of Maier's remarkable productions may
be sought in the Basilian emblems and also notably in
certain alchemical designs occurring in the works of
Andreas Libavius (*d.* 1616).   For this reason, an account
of some of the latter designs forms a convenient approach
to a discussion of Maier's work.   Libavius' fame rests
chiefly upon his celebrated *Alchymia*.   First published in
1597, this work is essentially a text-book, giving a clear
and comprehensive survey of the genuine chemical knowledge of the time.   It includes a number of new practical
observations of considerable importance, made by Libavius
himself; and it abounds in workmanlike descriptions and
drawings, not of apparatus only, but also of Libavius'
ideal ' chemical house ', or laboratory, which he delineates
both in plan and elevation.   ' Besides the main laboratory,' remarks Holmyard,[1] this contained ' a store-room
for chemicals, a preparation room, a room for the laboratory assistants, a room for crystallisation and freezing, a
room for sand and water baths, a fuel room, and, not least
among the amenities, a wine cellar—a delightful feature
unhappily overlooked by the modern architect of chemical
laboratories! '   In this work, Libavius strikes a reverber-
ating practical note; but, on the whole, like his apparent
contemporary, Basil Valentine, he displays a curious blend
of sound chemical knowledge with wild alchemical imagin-

ing. The works of Libavius and Basilius create, indeed, the impression that the long repressed fires of the true chemistry were at last beginning to break through the solid alchemical crust which had confined them for so long.

Libavius was a prolific writer on alchemy and medicine. Also, like Maier and Fludd (p. 229), he entered into the Rosicrucian controversy, originally on the opposite side, as a hostile critic of the supposed brotherhood. For a time he was professor of history and poetry at the University of Jena, and later he taught at Rothenburg-ob-der-Tauber, of which picturesque old city he was also chief physician.[2]

There is a copy in the St. Andrews collection of an enormous volume composed of writings of Andreas Libavius, published ' by special grace and privilege of the Emperor ' at Frankfurt in 1606 by Johann Saurius. Bound in the original stamped pigskin, and carrying still the remains of ancient thongs, this truly monumental alchemical work measures about 9 by 13½ inches and weighs some 10 pounds. In the forefront comes the famous *Alchymia*, followed by the somewhat similar *Commentariorum Alchymiae* . . . (' Handbook of Alchemy '), the second part of which includes a less practical and more mystical section on the Philosopher's Stone. This contains three curious and typical alchemical engravings which are of particular interest, since they are lettered and described in the text.

The drawings purport to illustrate ' the Form of the Art of Gold-making or Stone ' (*Typus Artis Aurificae seu Lapidis*), and of the first one (Fig. 15) it is said that ' Heinrich Kuhdorfer, in 1421 A.D. gave the whole of this illustration and added it to an ancient book in manuscript of the year 1028 '. A translation of the Latin description is given below:

' As we have no practice in the art of painting, we have produced at least the outlines, which each may fill up at pleasure. The blanks are to be filled in as follows :

FIG. 15.—An Emblem of Heinrich Kuhdorfer.
From *Commentariorum Alchymiae*, Libavius, 1606.

' (A) A small charcoal fire under the glass.

' (B) This space is filled by a winged serpent with a long tail, wearing a crown like the fabulous basilisk. It lies on its back with its feet in the air, and bites its tail, which is doubled back, thus being the dragon which is said to eat its own tail. It is of horrible appearance, green in colour, with a grey or ashy tail.

' (C) Here an eagle is to be painted with saffron feet and beak, wings outspread, many-coloured plumage in the wings, body, and tail, some feathers being white, others black, green, yellow, as in a picture of a peacock's tail, or the rainbow.

' (D) In the field, a black crow. (E) A red rose on a silver field. (F) A white rose on a red field. (G) A maiden's head, silver, representing the Moon. (H) A lion's head, gold, representing the sun. (I) One red rose on a silver field. (K) A saffron-coloured candle. (L) Three gold stars on a silver field. (M) Six dark blue stars on a golden field. (N) A white or silver candle; may also be saffron-colour. (O) Three white roses on a red field. (P) A king holding red or blood-coloured lilies in his hand. (Q) The king's wife or mother holding white or silver lilies. (R) A silver candle. (S) A red candle. (T) A plate of a crown, gold. (V) Candles in an apple. (X) Gold and silver lilies.

' What each of these means is not easy to say, for their applications are different, and there is also occasional variety in the Work itself. By the fire under the glass, you may understand the element of fire; by the saffron, red, and white candles inside the glass, the hidden fire of the sun and moon; by the candles in the crown, another fire which they call that of the three threads, otherwise the heat of the sun.

' The serpent is the mercury of the philosophers, which elsewhere means volatility, also solution; but by the eating of its tail, coagulation and fixity. The black crow indicates the blackness of putrefaction, the many-coloured eagle the diverse colours which appear during

PLATE 51

The Stone is likened to Coral.

From *Atalanta Fugiens*, Maier, 1618. (See p. 244.)

PLATE 52

The Oratory and the Laboratory.

H. F. Vriese *pinxit*, Paullus van der Doort, Antwerp, *sculpsit*. From *Amphitheatrum Sapientiae Æternae*, Khunrath, 1609.

PLATE 53

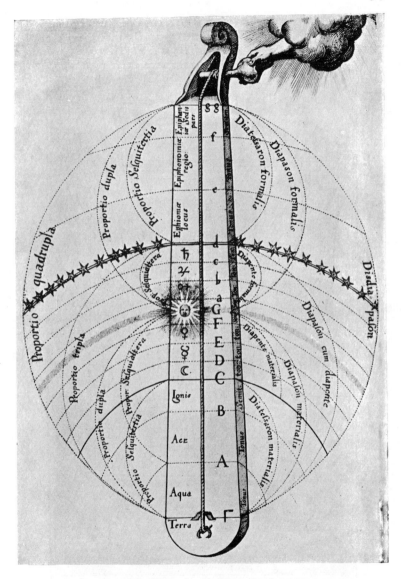

The Macrocosmic Monochord.

From *Utriusque Cosmi Maioris scilicet et Minoris Metaphysica, Physica atque Technica Historia*, Fludd, 1617.

PLATE 54

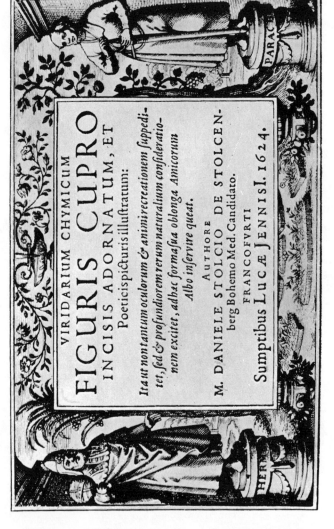

Title-page of *Viridarium Chymicum*, 1624. (See p. 256.)

fermentation. The different roses in different fields signify the heteromorphic nature of shining eggs. For the red stone can be produced as well out of the red as the white, and so can the white stone. The same applies to the king and queen, lion and virgin, which also signify male and female matter. The six dark blue stars are thought to typify the same number of parts of the mercury; the three red or golden ones, the same number of portions of the body to be dissolved.'

It is seen from this description of the early seventeenth century that the drawing incorporates many of the common alchemical conceptions which have been mentioned earlier in these pages, notably in the account of the Philosopher's Stone. The circle, for example, represents the ' shining ' Philosopher's Egg (p. 149), and the three enclosed symbols depict various operations of the Great Work.

The second drawing (Plate 40) is more elaborate, and depicts the Philosopher's Egg in several guises. It incorporates a striking variety of alchemical imagery, and for this reason has been reproduced by Diderot,[3] de Givry,[4] and others. Libavius calls it ' Another sketch of the Philosophic Work ', and describes it in this wise:

' (A) A pediment or foundation, like the earth.

' (B) Two giants or Atlases, resting on the foundation, who, to the left and right, hold up a globe on their shoulders and prop it with their hands.

' (C) A four-headed dragon, breathing forth, upwards towards the globe, four stages of fire: from one mouth there must come as it were air; from the second, thin smoke; from the third, smoke with fire; from the fourth, pure fire.

' (D) Mercury with a silver chain; beside him, two beasts bound with the chain and crouching. (E) A green lion. (F) A dragon, crowned, single-headed. These two beasts mean the same thing, namely, the mercurial liquid which is the first matter of the stone.

' (G) A three-headed silver eagle, two heads drooping and as it were fading, while from the third it must pour forth white water or the mercurial liquid, into the place of the sea (marked H).   (I) The picture of a wind (god), blowing downwards to the sea beneath.   (K) The picture of a red lion, from whose breast is to flow red blood into the sea beneath, etc., which must be so coloured as to seem a mixture of gold and silver, or of red and white. This symbol fits body, soul, and spirit, according to those who require three things even in the beginning; or it will fit the blood of the lion and the gluten of the eagle.   For those who assume three, have a double mercury; while those who assume two have one (mercury) only, which is reduced from the crystal or impure metal of the philosophers.

' (L) An expanse of black water, as in chaos: there is signified (thereby) putrefaction.   From this emerges a sort of mountain, black at the base, white at the summit, with an overflow of white streaming down from the summit.   For it is the symbol of the first solution and coagulation, and again of the second solution.   (M) The aforesaid mountain.   (N, N) Black heads of crows looking out from the sea.   (O) A silver rain falling from clouds on to the summit of the mountain, whereby is figured, firstly, the nutrition and ablution of Lato by Azoth; secondly, the second solution whereby the element of air is brought forth from earth and water.   Earth is the appearance of the mountain; water is that first liquid of the sea.   (P) The shape of clouds, from which comes dew, or rain, and the nutritive liquid.   (Q) The form of the sky, in which must lie a dragon, on his back, eating his own tail; this is the image of the second coagulation.

' (R, R) A negro and negress, holding up two globes, above and to the side of them.   They are supported by a larger globe, and signify the blackness of the second operation, in the second putrefaction.   (S) Here let there be represented a sea of pure silver, which typifies the mercurial liquid, the medium by which the tinctures are connected. (T) Here paint a swan swimming on the sea and spewing out of its mouth a white fluid. This swan is the white elixir, the white chalk, the arsenic of the philosophers, common to both ferments. It should support on its back and wings a globe placed above it.

'(V) An eclipse of the sun. (X, X) The sun rising from the sea, that is from the mercurial water, into which also the elixir must pass. Thereupon comes forth the real eclipse of the sun. On either side of it should be painted a rainbow, to indicate the Peacock's Tail, which then appears in the coagulation.

'(Y, Y, Y) An eclipse of the moon, which also should have a rainbow at the sides, and near the rainbow part of the sea for the moon to sink into; this is the symbol of the white fermentation. The sea, in both pictures, should be a little dark. (Z) The moon sinking into the sea.

'(b) A queen, decorated with a silver crown, stroking a white or silver eagle which perches beside her. (a) A king, in a purple robe, with a gold crown, having a lion standing beside him. In one hand the queen should carry a white, the king a red lily.

'(c) A phoenix perched on a globe and burning itself; from the ashes must fly a number of birds, silver and golden. For this is the symbol of augmentation and multiplication.'

Both figures thus depict the Vase of Hermes, heated at the base, in which occur in due sequence the various operations culminating in the production of the white and red Stone, symbolised by the Queen and King, respectively. 'The nutrition and ablution of Lato by Azoth' is a familiar alchemical conception signifying the transition from the black (putrefactive) to the white stage of the Great Work (p. 162).

The third drawing (Fig. 16) is of a simpler kind; it is found repeatedly, in various forms, in alchemical writings. The double-bodied lion represents the two principles, sulphur and mercury, sometimes called 'the two sulphurs', or 'the two mercuries'. The Bath of the

Philosophers, shown at the top, is described by Artephius [5] as ' the most pleasant, faire, and cleere Fountaine,
prepared onely for the *King* & *Queene*, whom it knoweth
very well, and they know it; for it drawes them to it selfe,
and they abide therein to wash themselves two or three
dayes, that is, two or three *moneths*; and it maketh them

FIG. 16.—The Bath of the Philosophers.
From *Commentariorum Alchymiae*, Libavius, 1606.

young againe, & faire. . . . Let therefore the *spirit* of
our living water, be with great wit and subtilty fixed with
the *Sunne* and the *Moone*, because they being turned into
the nature of water, doe dye, & seeme like unto the dead;
yet afterward being inspired from thence, they live,
encrease, and multiply like all other *vegetable* things.'
Libavius gives the following description of this drawing:

' (A) A double (-bodied) lion with one head, whereby is signified the first matter of the stone, from double mercury, which is leonine. The head spews out green water, that is, the mercury of the philosophers, engendered of both. This is also called the green lion.

' (B) Lions, on both sides, as if on the stair of Solomon, five in all, to signify five metals from one root, the leonine, and roasted mercury. These can pass into sun and moon: the lions on the right tend to the sun, those on the left to the moon; and this is seen in the operation by the succession of powers and permutation. It is not without a secret significance that the top lion is looking down at the lower ones, etc.

' (C) A picture of the sun. (D) A picture of the moon.

' (E) A bath, in which sits a king with a queen. This is also the type of the marriage-bed, for the pro-creation of their kind. Also a garden with a tree in it, bearing the apples of the Hesperides. (F) A king with a crown and sceptre decorated with lilies; he seems to address the queen. (G) In the middle, a tree bears golden fruits, while golden stars surround crowns, to signify multiplication and increase, or else the fruit of projection.'

This emblem may be interpreted as follows. Before entering the Bath or Fountain of the Philosophers, the King and Queen divest themselves of their garments.   In a like manner, gold and silver are divested of their im-purities in the preparation of sophic sulphur and sophic mercury (p. 132). The Bath of the Philosophers, like the First Key of Basil Valentine (p. 202), is thus a symbol of the purification of gold and silver, for the purposes of the Great Work.

§ 3. THE TWELVE CHOSEN HEROES OF CHYMISTRY

A prominent part in issuing works adorned with illus-trations which we have called ' the Frankfurt Emblems '

(p. 213) was taken by the publishing firm of Lucas Jennis. One of the earliest publications of the kind to emanate from this firm was the *Tripus Aureus* of 1618 (p. 169), with the famous illustrations of the Twelve Keys of Basil Valentine; later, in 1625, came the *Musaeum Hermeticum*, noteworthy for its engraved title-pages and the quaint plates in the *Book of Lambspring* (p. 28).

The year preceding the publication of the *Tripus Aureus* witnessed the appearance of one of the most notable works of its editor, Michael Maier. At Frankfurt, in 1617, there was printed by Anthony Humm, at the cost and charges of Lucas Jennis, a massive work of more than six hundred pages, written in Latin by this author under the title *Symbola Aureae Mensae Duodecim Nationum*. A preliminary idea of its nature may be gathered from the following translation of the title-page: ' Contributions of Twelve Nations to the Golden Table; that is, Hermæa or the Feast of Mercury, celebrated by twelve chosen Heroes, equal in their chemical experience, their wisdom and authority; intended for the confutation and disarming of Pyrgopolynices, or that Adversary who has for so many years braggingly assailed the maiden Chymistry both with fallacious arguments and pointed abuse, and for the restoration to their own honour and glory of Artists who have deserved well of Her:

' Wherein the continuity and truth of the Art is shown by 36 reasons and by experience, and more than three hundred books of the authorities:

' A work most useful not only for Chymistry but also to all others who are very eager after Antiquity and matters well worthy of being known; explained and set forth in 12 books, illustrated with copper-plates.

' By Michael Maier, Count . . . formerly of the Court of His Imperial Majesty.'

The handsome title-page has an ornamented border enclosing medallion portraits of the twelve chosen Heroes.[6] These are, in order: Hermes the Egyptian, Maria the Jewess, Democritus the Grecian, Morienus the

Roman, Avicenna the Arabian, Albertus Magnus the German, Arnoldus Villanovanus the Frenchman, Thomas Aquinas the Italian, Raymond Lully the Spaniard, Roger Bacon the Englishman, Melchior Cibiensis the Hungarian, and Anonymus Sarmata—the last of whom is to be identified with Michael Sendivogius the Pole.

Pyrgopolynices, the central figure of Plautus' ' Braggart Soldier ', is the Latin equivalent of Captain Bobadil, the swaggering boaster in Ben Jonson's ' Every Man in his Humour '. One by one, in Maier's lengthy work, he encounters the twelve chemical champions; and the ensuing debates, conducted after the manner of the schoolmen, form the main substance of the twelve sections of the book. At the beginning of each section there is a copper-plate engraving [7] which depicts the champion in a symbolical setting. Thus, Hermes (Plate 41 (i)) holds an armillary sphere in one hand and points with the other to a representation of Sol and Luna surrounded by fire; Avicenna watches the attempts of an eagle (sophic mercury) to fly away from a toad (sophic sulphur) to which it is chained (Plate 2 (ii)); Albertus Magnus admonishes a hermaphroditic figure; Thomas Aquinas watches the generation of metals from sulphur and mercury (Plate 3); and Roger Bacon [8] balances the elements, fire and water (Plate 41 (ii)), and awaits ' welcome gifts ' as the result of his skill (p. 259).

This is the description of Pyrgopolynices, ' adversary of Chymistry and Chymists ', as he takes the field against Hermes:

' And now the enemy stood in the midst. Burning with madness, and swollen and boastful with false persuasions, he thought little of all present compared to himself, and declared that with one little puff he would blow away and lay flat books past all counting. Presently he said: " Lo now, I will give a first taste of my power in your person, since you are the champion of the rest, and go about to defend and do battle, along with your allies, on behalf of that miserable damsel Chymistry, who is totally detestable and accursed to me ".   And this is the way of it:

' *First Argument of the opposers of the truth of Chymistry*

' No kind is capable of changing into another kind.

' But lead, copper, and other such metals are kinds in themselves, and so is gold in itself.

' Therefore neither are lead, copper, nor other such metals capable of being changed into gold.

' *Reply of Hermes*

' If kind could not be changed into kind, nor an in-dividual of one kind into an individual of another, then silver could not be naturally produced from lead, nor gold from silver.

' But this does happen, though not so quickly as hens' eggs are hatched and changed into chickens, as may be learned from miners and from daily experience.   There-fore, etc.'

The first refutation of Hermes is thus based upon the mediaeval belief that gold and silver are produced in nature from the metals with which we now know them to be associated in ores (p. 121); the origin of the noble metals is likened also to the generation of living organisms (p. 94).   Two further arguments put forward by the assailant of Chymistry are countered in turn by Hermes, who then advances his first affirmative reasoning on behalf of the ' miserable damsel ':

' If silver is produced naturally from lead, and gold from silver, there is no reason why gold should not be so produced by art, although not in the same way and manner, but in proportion to possibility and our power.

' But the first proposition is true.

' Therefore the second is true also.'

The pugnacious Pyrgopolynices fails to detect the false premise which the Egyptian champion uses in offence as in defence.   He allows three successive affirma-tive arguments to pass unchallenged, and leaves the triumphant Hermes in possession of the field. Nevertheless, the Father of Chymistry fails to show the imagination and resource which one would expect from the redoubtable author of the Emerald Table and of 35,999 other works on the ' Divine Art '!

With the advent of Maria the Jewess, Pyrgopolynices returns to the attack with his fourth, fifth, and sixth arguments against Chymistry. Maria (Plate 2) rebuts them to her complete satisfaction, and then puts forward the fourth, fifth, and sixth affirmative arguments. As before, the Adversary remains silent at this stage. His attitude is perhaps induced by gallantry towards the earliest woman chemist; for when Democritus follows upon her heels, Pyrgopolynices returns to the fray with all his original vigour.

So the arguments go on interminably, after the manner of alchemical literature, as if a prolific discharge of verbiage could create and establish facts, as well as ' lay flat books past all counting '. The twelve chosen Heroes advance with monotonous regularity and inflict their strokes upon this alchemical whipping-boy. Naturally, therefore, in his encounter with the tenth champion, Roger Bacon the Englishman (Plate 41 (ii)), the Adversary begins to show signs of exhaustion. However, ' recovering his self-possession ', he opposed Bacon as follows :

' *Twenty-eighth negative argument of Pyrgopolynices against Chymistry and Roger Bacon*

' If the imperfect metals should be transmuted into the perfect by nature, and if art should imitate nature in that matter; then the artist must live as long as nature, that is, over a thousand years.

' But the artist cannot live so long, for his entire life scarcely attains to 100 years.

' Therefore he cannot imitate nature in transmuting the imperfect into the perfect.

' *Reply of Bacon to the twenty-eighth negative argument*

' *I reply to your major premise*: that, if art could imitate nature in all things, then the premise would hold good. But art need not and cannot follow nature's footsteps in all things, since it chooses a shorter middle way from the same beginnings, borrowed from nature, to the same end. Hence it will not need so much time as nature, but far less. Nature's time is extremely long, and the fashion of her concoction is uniform, and her fire very slow. That of art, on the other hand, is short; the heating is controlled by the wit of the artist, as the fire also is made intenser or milder. Moreover, art mingles two or more things one with another, by the hands of the artist, which nature produced in separate places and could never have joined in one, since she lacks the power of motion in space, of separation and of choice; all of which the artist possesses; for he brings together from different mines certain things; he blends the homogeneous; he separates the heterogeneous; he makes critical choice of fires, vessels, colours and time, of maturity and all other circumstances, until he reaches the goal he desires: Hence it is that nature never produces perfect tinctures in the abstract, but only in the concrete; for instance, she never produces spirit of wine quite pure, in the abstract, but only in the concrete, that is in wine; from which it is separated out by art.

'*I reply to your minor premise*: that the artist does not need 100 years for the perfection of his art, much less a thousand, since a few days, not amounting in all to so much as a whole year, may be enough for him, once his art is perfectly known and discovered. But for its discovery, unless one has learned it from another, many years are needed, and in most cases even one's whole life has not been sufficient for its perfection. Ulysses spent 10 years wandering about various seas on his way home from Troy, which journey he could have accomplished within 20 days if he had wished. In like manner, some have sailed around the whole world in a few years' time, which a hundred years ago, before Magellan, no man could have accomplished in his whole life. The reason for this is that it is easier to follow a way once found than to discover one not yet found; so in art, Ulysses is the artist who errs in divers ways until he reaches the desired goal; but this once found, it is not hard to finish this great work in a few days.'

In such ways Michael Maier calmly and comfortably sets up and knocks down his ninepins throughout this ponderous work. To be quite fair to him, although his arguments are long and windy, they are nevertheless interspersed with many shrewd observations and gleams of wisdom. ' In art [*i.e.* the practice of chemistry], Ulysses is the artist who errs in divers ways until he reaches the desired goal ', is a dictum which would form a good motto to adorn the bench of a modern research chemist; or, again, the tail-end of Bacon's reply to the major premise contains a graphic thumb-nail sketch of the efficient chemist—who, in Maier's words, ' makes critical choice of fires, vessels, colours and time, of maturity and all other circumstances, until he reaches the goal he desires '.

Bacon's reply to the minor premise contains an ingenious allusion to his belief that one could sail westwards to reach the east, and it will be noticed that Maier endows the thirteenth-century sage with the posthumous knowledge, which would have so delighted him in life, of the realisation of this idea. Each of the twelve protagonists is selected as typical of his nation, and his arguments are supported by references to eminent fellow-countrymen, real or imaginary, who held a reputation for alchemy in Maier's day. Among the Englishmen, for example, are Hortulanus (alias Garlandus), Johannes Dastinus, Richardus Anglicus, Georgius Ripleus, Thomas Nortonus, Cremerus, and Eduardus Kelleus. Michael Scotus is mentioned also as celebrated in his own country for this art. This part of the work has a certain value, as it forms a series of early contributions to the history of alchemy; these are not, however, of a critical nature.

There is no need to follow the ' arguments ' further.

As a rule, they do not offer even the attraction of novelty; some of them, indeed, are taken almost verbatim from the second part of the first book of the *Summa Perfectionis* (p. 49) and other well-worn treatises. Suffice it to say that the author performs his task conscientiously, and that his exhausting work drags on through more than six hundred quarto pages to its appointed end, leaving the *advocatus diaboli* well and truly buried in a typical alchemical wood beneath a monumental mass of alchemical verbiage. There we are glad to leave him.

§4. COUNT MICHAEL MAIER

Michael Maier, whom we have encountered as editor of the *Tripus Aureus* and author of *Symbola Aureae Mensae*, was a remarkable man. He was born in 1568 at Rendsburg, a town in Holstein about nineteen miles west of Kiel, situated to the north of the present canal. A portrait (Plate 42) which appears in his *Symbola Aureae Mensae*, *Tripus Aureus*, and *Atalanta Fugiens* shows him as he appeared in the year 1617, at the age of forty-nine: he is attired as a noble, carrying a book and wearing a sword. A noteworthy detail of the accompanying crest is the familiar alchemical device of the toad of earth joined to the flying eagle by means of a chain (p. 223). A Latin inscription at the foot of the portrait states: ' The school gave me three titles, the Emperor other three; this remains for me, to be able to live well and die well in Christ '. Maier is here described as Count of the Imperial Consistory, Doctor of Philosophy and Medicine, sometime Physician to the Emperor, etc.

This portrait summarises a good deal of our knowledge of Maier's career; for, in spite of the eminence which he clearly attained among his contemporaries, the records of his life are curiously scanty. The only biography of Maier appears to be a short work by J. B. Craven,[9] published at Kirkwall in 1910; even this is out of print and almost as difficult of access as some of Maier's own works.   The details of Maier's degrees are not clear, but he is said [10] to have graduated as doctor of medicine at Rostock in 1597.   According to Craven, he left Holstein in 1608, at the age of forty.   His reputation must have been established by that time, for soon afterwards he was appointed physician to the Emperor Rudolph II at Prague: the Emperor raised him to the rank of Pfalzgraf, or Count Palatine, and made him his private secretary.   Prague was in those days a celebrated centre of alchemical activities (p. 22), the Emperor being an enthusiastic patron of ' the Divine Art '.   Maier was an omnivorous reader; and here, with the unrivalled library of the Emperor at his disposal, he seems to have laid the foundations of the numerous works which were to stream from his pen in rapid succession a few years later.

The Emperor Rudolph II died in 1612.   Soon afterwards Maier visited England; but if this was in 1614, as usually surmised, his visit was a long one, for the dedication of *Lusus Serius* is dated: ' Frankfurt-on-Main, immediately on my return from England and on the eve of my departure for Prague, September 1616 '.   He undertook many journeys, especially in Germany, during the course of which he lent an all too credulous ear to current myths and marvels in alchemical and other fields.   Elias Ashmole states that he visited England in order to learn English, so as to translate Norton's *Ordinall* into Latin verse, ' which most judiciously and learnedly he did ' (p. 174).   In England he may have met Robert Fludd, the philosophical mystic (p. 252).   Since his wavering faith in the marvellous Orcadian tree-bird was restored by ' a certain Scot, a doctor of medicine ' (p. 233), it has been surmised that Maier had a Scots correspondent. Two graduates of King's College, Aberdeen, each of whom, like Maier, studied at Rostock, have been proposed for this distinction.[11]   The first of them, Duncan Liddel,[12] has been described as one of the first physicians and philosophers of his time; he became professor of mathematics, and, later, professor of medicine, at the

University of Helmstedt. He returned to Aberdeen in 1607, and died there in 1613. Dr. John Johnston,[13] the second candidate, was at Rostock in 1584; he died at St. Andrews in 1611.

In 1619 Maier was appointed physician to the Landgrave Moritz of Hesse. From 1612 until his death in 1622 he was domiciled at Magdeburg,[14] where he practised as a physician. It appears likely that valuable records of Maier's life and activities were destroyed in the sack of Magdeburg by Tilly in 1631.[15] In religion, Maier was a Lutheran.

Such are the fragmentary details of the life of one of the most versatile and picturesque figures in the long and romantic history of alchemy. Maier was not only an alchemist; he was also a philosopher, a mystic, an exponent of Rosicrucianism, a physician, a classical scholar, and a musician. In alchemy, Maier was a writer, rather than a laboratory worker, although he can scarcely have escaped the prevailing vogue of experimenting in search of the Philosopher's Stone. His alchemical works are of the mystical allegorical type, without the slightest pretension to practical value.

Upon turning to these writings, it is astonishing to find that Maier's first book was published in 1614, when he was forty-six, and that the whole of his voluminous output appeared within the succeeding six or seven years. This first work, *Arcana Arcanissima* (' The Most Secret of Secrets '), was dedicated to Sir William Paddy, physician to James I of England, and fellow of St. John's College, Cambridge. It consists of a discourse in six books, devoted to an alchemical interpretation of Egyptian and Greek mythology, a dominating theme in Maier's alchemical writings. In 1616 he published *De Circulo Physico Quadrato* (' On Squaring the Physicist's Circle ') and *Lusus Serius* (' Serious Pastime '); and in 1617 at least four of his works appeared. These were: *Symbola Aureae Mensae Duodecim Nationum* (' Contributions of Twelve Nations to the Golden Table '), *Jocus Severus*

(' A Jest in Earnest '), *Examen Fucorum Pseudo-Chymicorum Detectorum* (' A Swarm of Pseudo-Chymical Drones Exposed '), and *Silentium Post Clamores* (' Peace after Noise '). *Themis Aurea* (' Golden Themis '—Themis being the goddess of justice), *Tripus Aureus* (' The Golden Tripod '), *Atalanta Fugiens* (' Atalanta Fleeing '), and *Viatorium* (' The Traveller's Companion ') were his chief publications in 1618 (see Plates 44 and 13). In 1619 appeared *Verum Inventum*, otherwise *Munera Germaniae* (' A True Discovery, or Germany's Gifts '); and in 1620, *Septimana Philosophica* (' The Philosopher's Week ').

The *Symbola Aureae Mensae*, which has already been reviewed (p. 222), is the largest of these works. This has been characterised aptly as ' interesting, though perhaps rather exhausting from its bulk '; less aptly, as ' full of the results of researches '.[16] The most interesting of Maier's publications is *Atalanta Fugiens*: this is described below. Some of his productions, such as *Lusus Serius* (p. 257) and *Jocus Severus*, are trivial: the latter work is concerned with a parliament of birds, in which, after a wordy debate, the owl is made queen and crowned.

The active part which Maier took in publishing the most celebrated of the Basilian writings (p. 196) suggests that he would have known their author if the latter had been a contemporary writer. Nothing in Maier's writings supports this view, or throws any light upon the identity of Basil Valentine.

Maier was an early exponent of the Society of Rosicrucians, or Brethren of the Rosy Cross, an ill-defined body of sectaries who appear to have combined the alchemical code with their peculiar system of mystical philosophy. There is no conclusive evidence that he was a member of the Fraternity, although he advocated its cause in some of his writings, notably in *Silentium Post Clamores* and *Themis Aurea*. In *Symbola Aureae Mensae* he states on p. 290: ' The fame of the aforesaid Fraternity, which here fills many men's ears and employs their tongues, . . . reached me while I was in England, devoting my attention wholly to chymistry. As it was brought by obscure gossip, incredible and grossly exaggerated, I gave it no more belief than was due to the credibility of the reporter. . . . But when a book dealing with their fame and confession appeared (at Frankfurt in the autumn of 1616), I happened to read it, and was brought to have a very different opinion concerning them . . . since the Rosicrucians put forward nothing contrary to true piety, to nature, to any condition of men, nor finally, to virtue or justice, but every detail aims at the praise of the Creator, the revelation of nature and the well-being of God's creatures, we shall with duteous and well-deserved prayers await from this laudable order whatever will come about.'

It is interesting that Sir Isaac Newton, who made extensive manuscript transcripts from Maier's alchemical books, referred to this entry in a note which he made on the fly-leaf of his copy of Thomas Vaughan's *Fame and Confession of the Fraternity of R : C : commonly, of the Rosie Cross* (1652). The note mentions that 'R. C. the founder of ye supposed Rosy Crucian society (as the story goes) was born anno 1378 dyed anno 1484 ', and directs attention to the passage by Michael Maierus ' in his symbola aurea mensa dated in December 1616 where (pag 290) he notes that ye book of Fame & confession were printed at Francford in autumn 1616. This was the history of y^t imposture.' Newton's note reflects the uncertainty concerning the existence of the alleged Society and the reliability of the literature dealing with it.

Maier was a prodigal writer on alchemy, and his works achieved a high reputation in the unsophisticated atmosphere of the early seventeenth century. As Waite[17] remarks: ' Maier is distinguished from his peers in German alchemy by an extraordinary obscurity of style and by a passion for extensive titles '. He showed no critical faculty in his writings, and his credulity was unbounded. For example, his *Tractatus de Volucri Arborea* (' A Treatise concerning the Tree-Bird '), published by

PLATE 55

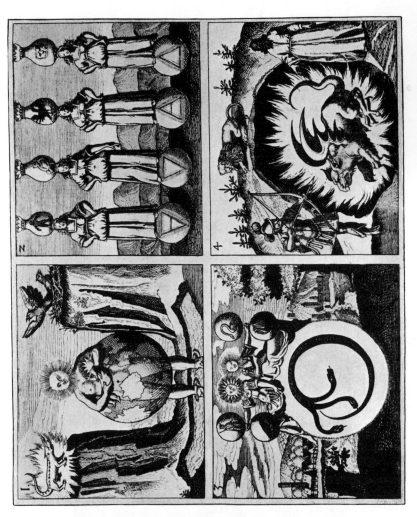

The First Four Emblems of Mylius.

From *Philosophia Reformata*, Mylius, 1622. (1) The Earth is the Nurse of the Stone.
(2) The Four Sisters. (3) The Spirit, Soul and Body of the Stone. (4) The Lion
and Dragon (Sulphur and Mercury). (See pp. 261, 262.)

PLATE 56

(i) Sublimation.

From *Viridarium Chymicum*, Stolcius, 1624 (reprinted from
Philosophia Reformata, Mylius, 1622).

(ii) Multiplication.

From the same. (See p. 264.)

PLATE 57

(i) The Two Sulphurs.

From *Viridarium Chymicum*, Stolcius, 1624 (reprinted from
Philosophia Reformata, Mylius, 1622). (See p. 265.)

(ii) The Seven Metals.

From the same. (See p. 265.)

PLATE 58

The Tree of Universal Matter, with Senior and Adolphus.

From *Philosophia Reformata*, Mylius, 1622. (See pp. 267, 269.)

Lucas Jennis at Frankfurt in 1619, is devoted to a solemn and long-winded disquisition ' on the origin, miraculous rather than natural, of vegetables, animals, men, and certain supernatural creatures', including in particular ' the Tree-Bird, without father or mother, in the Orkney Isles, in the prevailing shape of goslings '.   In this book of mare's-nests Maier swallows without difficulty the fable of the geese which were said to hatch out of barnacles— ' that marvellous bird ', as he calls it, ' which, in the Orkney Isles lying in the ocean beyond Scotland, is born in trees, or rather logs '.   Although he admits that ' many are in doubt about the existence of this bird, or whether indeed it arises from Sicilian nonsense [*ex gerris Siculis*] ', he affirms that ' lately some of those goslings have been pointed out to me by a certain Scot, a doctor of medicine, lying in their own shells . . . in which more than fifty were sticking to the log.   Hence I have taken the opportunity of recording the description of so rare and wonderful a birth (which none before* has dealt with methodically, except in passing), in this treatise wherein we shall touch on many marvellous matters of the same kind arising contrary to the course of nature, with the intent that the glory of God may be enhanced even by this.'

In another place Maier affirmed that Albertus Magnus possessed ' a stone naturally marked with a serpent, and endowed with so admirable a virtue that on being set down in a place infested with such reptiles, it would attract them from their hiding places.' [19]   His activities in mystical alchemy have been termed ' ruinous follies ', to which he sacrificed his time, fortune, reputation and health.[20]   Undoubtedly Maier wrote much portentous nonsense, but a great deal of it was at the same time erudite nonsense.   He shows all the marks of the widely-read scholar of his day, and his writings abound in allu-

---

* Sir John Mandeville (c. 1360), among others, had already encountered this *rara avis*, if the celebrated *Travels* are to be trusted: ' For I tolde hem ', writes the narrator, ' that in oure Contree weren Trees, that baren a Fruyt that becomen Birddes fleeynge; and tho that fellen into the water lyven; and thei that fallen on the earthe dyen anon; and thei ben right gode to Mannes mete.' [18]

sions to classical literature, such as his references in the preface of the *Tractatus de Volucri Arborea* to ' Sicilian nonsense ' and to ' the musician who is laughed at when he always blunders at the same chord '.[21]

It was probably his extensive knowledge of classical literature which led Maier to attempt a detailed correlation of alchemy with the mythology of Greece and Rome, and to a less extent with that of Egypt. Up to the fifteenth century, alchemical writers had used mythological fables only to a slight extent as an allegorical cloak for their supposed priceless secrets. This practice, however, was carried to great lengths by certain post-Renaissance writers, of whom Maier is the most noteworthy. He began to apply mythology to alchemy in his first book, *Arcana Arcanissima*, and developed the idea more fully in *Atalanta Fugiens*. This system of Hermetic mythology, so characteristic of the most extravagant phase of alchemy, reached its apogee late in the eighteenth century in the writings of Pernety (p. 245), who adopted unreservedly the startling and untenable thesis that classical mythology was a vast web spun by ancient writers for the express purpose of enmeshing and concealing the secrets of the Hermetic Art in general and of the Great Work in particular.

At the present day, the appeal of Maier's works is due largely to the symbolical illustrations by Johannes Theodorus de Bry and his associates which adorn some of them, including, in particular, *Atalanta Fugiens* and *Symbola Aureae Mensae*.[22] A reference to the de Bry family is therefore of interest at this point.

Johannes Theodorus de Bry (1561–1623) collaborated in much of his earlier work with his father, Dirk (or Theodorus) de Bry (1528–1598), and his younger brother, Johannes Israel de Bry (*d.* 1611).[23] About 1570 de Bry senior, having embraced the reformed religion, was forced to migrate from Liége. Between 1585 and 1590 he settled at Frankfort-on-Main, where he founded a business as a seller of prints and books, besides which he published many large works illustrated by himself and his two sons. His copper engravings were marked by great spirit and expression, but he was surpassed in technical skill and design by his elder son, Johannes Theodorus. In 1587 de Bry senior visited England, and made the acquaintance of Richard Hakluyt, who provided him with material for the illustrations of sundry books of voyages and travels.

Notable among these was *A briefe and true report of the new found land of Virginia*, by Thomas Hariot, describing the first English settlement in Virginia, on the island of Roanoke, lasting from 1585 to 1586. This work, first published in 1588, was reissued by de Bry at Frankfurt in 1590, with copper engravings founded upon the original drawings of John White. In de Bry's words, it contained ' the true Pictures of those people wich by the helpe of Maister Richard Hakluyt of Oxford Minister of Gods Word, who first Incouraged me to publish the Worke, I creaued out of the verye original of Maister Ihon White an Englisch paynter who was sent into the contrye by the queens Maiestye, onlye to draw the description of the place, lyuely to describe the shapes of the Inhabitants their apparell, manners of Liuinge, and fashions. . . . I craeued . . . them at London, and brought Them hither to Franckfurt, wher I and my sonnes hauen taken ernest paynes in grauinge the pictures ther of in Copper ' [24] (Plate 43).

There can be little doubt that Shakespeare drew upon this material in writing *The Tempest*, the only play in which he leaves Europe and the Mediterranean: [25] Caliban, in particular, the American savage depicted from the contemporary literature of American exploration, seems to stand revealed in John White's drawings and the de Brys' derived engravings.

Johannes Theodorus de Bry continued his activities at Frankfurt after the deaths of his father and brother, which occurred before the publication of Michael Maier's works. Upon his own death, the de Bry publications were continued by his two sons-in-law, Matthew Merian and William Fitzer. The original plates of many of the engravings were preserved long after his day and used in later works, as for example in the 1687 edition of *Atalanta Fugiens*, issued under the title *Secretioris Naturae Secretorum Scrutinium Chymicum* (p. 238).[26]

§ 5. ATALANTA FLEEING

We come now to the most interesting, curious, and rare of the many works of Michael Maier. This, by reason of a feature which is certainly arresting and probably unique, has led us to apply the designation of ' A Musical Alchemist' to this versatile physician, philosopher, alchemist, and classical scholar. Maier's *Atalanta Fugiens*, or ' Atalanta Fleeing', was printed at Oppenheim by Hieronymus Gallerus, at the cost and charges of John Theodore de Bry, in the same year (1618) which saw the appearance of the *Tripus Aureus* at Frankfurt. This remarkable work contains fifty copper-engravings of emblematic pictures, designed largely to illustrate the supposed relationships between alchemical doctrines and classical mythology. Each engraving is provided with a cryptic title, a discourse, and a Latin epigram written in elegiac couplets; the epigrams are set to music (presumably of Maier's own composition), in agreement with the belief held by many of the adepts that music also entered into the mysteries of the Hermetic Art (p. 247). A German version of each of the epigrams is also given.

The full title runs as follows: ' Atalanta Fleeing, that is to say, New Emblems of the Secrets of Chymical Nature: Adapted partly to the eyes and partly to the understanding, in copper-plates, with the addition of sentiments, epigrams and notes; partly to the ears and the refreshment of the mind, in about 50 fugues for three voices, whereof two are fitted to one simple tune, well suited to singing couplets; the whole most excellently pleasant to see, read, meditate, understand, judge, sing and hear '.

The main theme of the handsome pictorial title-page (Plate 44) is the mythological story of Atalanta and the golden apples.[27]   According to this entertaining fable, Atalanta, being the fleetest-footed of mortals, had developed an effective technique for disposing of her suitors, by challenging them to race with her: the first suitor to win would gain her hand, but the losers were to be put to death.   She was finally beaten by the enterprising Hippomenes, with the help of Aphrodite (Venus), who had given him three golden apples to drop on the ground during the race.   Atalanta, ' like a girl, valuing the giddy pleasure of the eyes ', stooped to gather them; thereby lost the race; and, true to the rules of her game, became the wife of Hippomenes.   As the poet observes:

> The nimble Virgin, dazled to behold
> The glittering apple tumbling o'er the mold,
> Stop'd her career to seize the rowling gold.

There was an unfortunate post-nuptial incident.   The happy couple, having wandered into a temple of Cybele and profaned it by their embraces, were metamorphosed by that touchy goddess into lions.

These picturesque happenings are illustrated at the foot of the title-page of *Atalanta Fugiens*.   Above, on the right-hand side, Venus is depicted in the act of handing the golden apples to Hippomenes.   At the head of the engraving are shown Aegle, Arethusa, and Hesperia, the three guardians of the golden apples of the Hesperides, with the attendant dragon, Ladon.   Below them, Hercules gathers the golden fruit.

The musical accompaniments to the epigrams are canons in two parts against a repeated *canto fermo*, as explained in the Appendix below.   The three parts are quaintly termed, in allusion to the title-page: ' Atalanta, or the fleeing voice '; 'Hippomenes, or the pursuing voice '; and ' The apple in the path, or the delaying voice '.

The same collection of engravings was reprinted, without the music, at Frankfurt in 1687, under the title *Secretioris Naturae Secretorum Scrutinium Chymicum*: ' A Chymical Examination of the more secret parts of Nature, illustrated by most ingenious Emblems, carefully adapted to the eyes and understanding, most aptly set forth in copper-plates; and by adjacent sentences, remarkably in point, and also very learned Epigrams. A little Work much desired and longed for by the higher wits, born for nobler ends, on account of the themes treated in it, which are subtle, lofty, holy, rare, and generally exceedingly abstruse; Now for the second time set forth for the benefit and profit of the most noble Republic of Chymistry, to be read, meditated, understood, and judged with extraordinary delight.'

In *Atalanta Fugiens* Maier handles the Philosopher's Stone in an extraordinary way, and, in general, draws the most extravagant parallels between alchemy and mythology. To the first two engravings he assigns titles taken from the Emerald Table of Hermes: ' The wind has carried him in its womb '; ' The earth is his nurse '. He continues, with apparent irrelevance: ' Go to the woman that washes clothes, and do thou likewise '. The epigram to the corresponding illustration (Plate 45) runs: ' Whoever thou art that lovest to examine hidden doctrine, do not idle, but take as thine example everything that can profit thee. Seest thou not, how a woman is wont to wash clothes clean of stains by pouring hot water upon them? Imitate her; so shalt thou not be cheated of thine art, for the wave washes the lees of the dark body.' A reference to (Plate 23 (ii)) shows that a variant of this illustration appeared in the *Viridarium Chymicum*, in 1624: alchemical ideas were limited in number, so that alchemical artists and writers were forced to seek novelty by means of slight alterations in their modes of expressing familiar tenets. Their emblematic drawings, like their expressions, often give rise to a feeling that they have been constructed from ' standard parts '.

The imagined similarity between the origin of metals and the process of growth in living organisms is brought out in an engraving of a husbandman scattering seed in a tilled field (Plate 14). In another place (Plate 17) there is a reference to the Philosopher's Egg.

The idea of the operation of masculine and feminine principles in the inanimate world is a leading feature of Maier's scheme. 'Make a circle of male and female,' he says, in introducing a curious geometrical design shown in one of the drawings (Plate 46), 'then a square, from that a triangle; make a circle, and thou shalt have the Philosopher's Stone.' This emblem is, in fact, substantially identical with Basil Valentine's synthetic symbol of the Great Work (Fig. 13, p. 207), and it emphasises the geometrical aspect of alchemy. In another place Maier repeats the alchemical *cliché* that the Sun needs the Moon as the cock needs the hen: the cock was sacred to the god Helios, or Sol. There are also hermaphroditic designs, one of which shows the 'Rebis', or 'Two-Thing', in an engraving which also depicts Mercury and Venus. 'The ancient adepts called a twin thing Rebis, because in one body it is both male and female, androgyne. For Hermaphroditus is said to have been born in twin mountains, he whom gentle Venus bore to Hermes. Despise not the doubtful sex, for unto thee this male that is likewise a female will give the King.' The two deities here mentioned are often associated in cult, and Ovid states that they are the parents of a bi-sexual godling, Hermaphroditus; this idea seems to be a Grecian variant of the bi-sexual gods of the Orient.[28] Sol (sophic sulphur) and Luna (sophic mercury), however, usually play the male and female parts in alchemical drama, and Maier shows them in the act of slaying the Dragon. In this emblem, as in so many others, the Dragon represents base matter, containing nevertheless the seed of gold; at the same time he is the guardian of the garden of the Hesperides and the fabled apples of gold. When he is killed, his seed is able to germinate and fructify (p. 95); the gate of the Hesperian garden is thrown open, and the golden apples may be plucked. It is usually stated that the Dragon must die in the company of his brother and sister, Sol and Luna, the particular function of Luna (sophic mercury) being to render him volatile and susceptible to change.

The production of colours in alchemical operations was always a source of wonder and admiration to the alchemists, who often likened transmutation to dyeing, or ' tingeing ', as we have already seen (p. 148). ' Three things are enough for the Magisterium,' writes Maier, ' white smoke, that is, water; the green lion, that is, Hermes' bronze; and stinking water.' All these expressions are synonyms for sophic mercury. Maier illustrates the whitening of Latona (p. 162), and symbolises in several ways the production of various colours in the operations of the Great Work. Thus, Venus runs to the slain Adonis and reddens the white roses with her blood as she goes. In another place the artist depicts a talking vulture, which perches upon the peak of a high mountain and cries without stint: ' White am I and black; citrine and red am I called, and I lie not at all '.[29]

There are several concealed references to the sulphur-mercury theory, and to the idea of fixed and volatile principles. In one picture a feathered Lion wrestles with a normal one; in another, two Eagles meet, one coming from the East and the other from the West; in a third, the Wolf from the East and the Dog from the West bite each other. The second of these three pictures may be allusive to Plutarch's statement that two eagles were sent, from opposite ends of the earth, to meet at Delphoi.[30] The third is strangely reminiscent of a passage [31] from the *Ts'an T'ung Ch'i* of Wei Po-yang, dating from 142 A.D. (p. 38): ' The Dragon breathes with the Tiger and the Tiger receives the spirit from the Dragon. They mutually inspire and benefit. They eat and devour one another.'

The death and resurrection of the King (gold) are.

represented by a wolf (antimony, or an acid) devouring the King; the wolf is then burnt (calcined), and the naïve picture shows the revivified King stepping away from the bonfire in jaunty abandon (Plate 47). This figure is clearly a variant of the First Key of Basil Valentine (p. 200).

Maier touches in a very interesting way upon the alchemical doctrine of the eternal nature and unity of all things. The first conception is embodied in the ancient device of a dragon biting its own tail: ' the Dragon . . . must be tamed with iron, with hunger, and with prison, until he swallows and vomits himself up again, kills himself, and bears himself '. This may be interpreted as a symbol of unceasing natural renewal, and of the imagined reversible transformation of matter which it was held that the alchemist should strive to imitate: the emblem is thus sometimes used to denote the Philosopher's Egg, or the circulation in it of the contained material. It is not inappropriate to remark here that it was a vision of the Gnostic symbol of a snake seizing hold of its own tail which led Kekulé, in 1865, to formulate one of the crowning achievements of chemistry, namely, the theory of the benzene ring. Truly, it may be said of chemistry as of Mary, Queen of Scots: ' En ma fin est mon commencement '.

The fundamental unity of all things is shown in a plate (Plate 48) depicting the four elements of Aristotle as four brothers who are threatened by an armed assailant. The epigram explains that ' four brothers stand in long array, whereof one holds up in his right hand the weight of earth, the second of water; the portion of the rest is that of air and fire. If thou wishest them to perish, overcome but one of them by death, and all will be taken off by kinsmen's dying, because the mutual bonds of Nature join them.'

In one of his most remarkable alchemical conceits Maier makes great play with the story of Oedipus, son of Laius, king of Thebes, and Jocasta. Oedipus, ignorant of his parentage, having slain Laius in a scuffle, solved the riddle of the Sphinx, succeeded to the kingship of Thebes, and married Jocasta. The city shown in the background of Maier's thirty-ninth emblem (Plate 49) is Thebes; the two mermaid-like creatures are the Sphinx, once barring the way into the city, and once about to throw herself from the rock after Oedipus has solved the riddle. The soldier threatening the king is Oedipus killing Laius. The marriage of Oedipus and Jocasta is also indicated. The three figures in the foreground, with characteristic marks on their foreheads, illustrate the answer to the Sphinx's riddle: ' What animal has four legs, two legs and three legs, and is weakest when it has most legs? ' Oedipus answered correctly: ' Man, who goes on all fours in infancy, walks upright when full-grown, and uses a staff when old and decrepit '. The following translation of Maier's epigram indicates a faulty knowledge of the story:

The Sphinx, who was terrible to Thebes by reason of her riddling
    speech,
  Oedipus by his craft hurled to her own death.
Her question was: ' What has twice two feet in the morning, but
    in the light
  Of noonday two, three when evening is come? '
Thereafter, being victorious, he slew Laius, who would not give
    place to him,
  And took to wife her who was his mother.

In his ' Discourse ' on this emblem and epigram, Maier airs his view that the stories of classical mythology are merely allegories embodying alchemical truths:
   ' Now the riddles of the Sphinx are said to have been very many, but this, which was proposed to Oedipus, was the chief one: four-footed in the morning, two-footed at noon, three-footed at eve—what is it? What Oedipus' answer was is unknown (!), but some make it signify the life of man, and they are wrong.
   ' For the square, or four elements, must first of all be

considered. From this we come to the hemisphere [semi-circle], with its two lines, the straight and the curved, in other words the white moon; and then to the triangle, composed of body, spirit and soul, or sun, moon and mercury. Wherefore Rhazes in his Epistle says *The Stone is triangular in essence, square in quality.*

'Emblem XXI [Plate 46] and its exposition are likewise to be referred to this. . . . [The story of Oedipus] was not written as sober history for anyone, nor as an example for his imitation. By philosophers, at all events, it was invented and brought in by way of allegory, to lay bare the secrets of their learning. . . For the first efficient cause, that is, the father, is put out of the way and routed by its own effect, that is, its son. . . . It [the effect] has swollen feet [*i.e.* Oedipus] because it cannot run, and is like a bear, as having a great secret, or a toad, which moves at a slow pace: because it is fixed, fixing another, and does not flee nor shrink from fire, which philosophers chiefly need, although it is a medium of little price'.

These references bear witness to Maier's capaciously extravagant imagination; but it is in his direct allusions to the Philosopher's Stone that he excels. He illustrates the supposed triune nature of the Stone by quoting the mythological story of 'three-fathered' Orion: [32] 'they say the offspring of Wisdom is thus three-fathered; for the Sun is said to be the first, and Fire the second father to it, and he that is excellent in the Art, the third'. He brings in also the story of Kronos (Saturn), who swallowed his offspring, with the exception of Zeus (Jupiter), for whom Rhea substituted a stone wrapped in swaddling clothes. [33] This tough morsel, known in Latin writings as Abadir, was — needless to say — the Philosopher's Stone! 'Wouldst thou know the reason why so many poets tell of Helicon,' asks Maier, 'and why its summit is the goal of each one? There is a Stone placed on the topmost height, a memorial, which his father swallowed and spewed up instead of Jupiter. If thou understand the matter just as the words sound [literally] thy mind is unskilful, for that Stone of Saturn's is the Chymists' Stone ' (Plate 50).

Descending from the mountain-top to the great waters, Maier likens the Stone in a later epigram to coral: ' As coral grows under the waters and is made hard in the air, so the Stone ', runs the title of the emblem (Plate 51). The epigram continues: ' A plant, flourishing moist beneath the waves of the Sicilian sea, has multiplied its branches beneath the warm waters . . . the name it goes by is CORAL . . . it becomes a stone . . . a ruddy colour it has; this is a fitting image for the Stone of Physick '.

With such an introduction, the reader is prepared for a still later picture (Plate 21) which shows the Stone to be of universal occurrence: ' The STONE is said to be vile refuse, and to lie by chance on the roads, that rich and poor may get it. In the high hills others declare [it is], among the breezes of the air, while others say it is nurtured by rivers. All these are true in their proper sense; but I ask thee to seek such great gifts in hilly places.' This engraving, like others in the work, conveys incidentally a charming impression of mediaeval scenes and activities. In order to illustrate Maier's literary style, the Latin [34] original of the above epigram is appended:

> Vile recrementum fertur LAPIS atque jacere
> Forte viis, sibi ut hinc dives inopsque parent.
> Montibus in summis alii statuere, per auras
> Aëris, at pasci per fluvios alii.
> Omnia vera suo sunt sensu, postulo sed te
> Munera montanis quaerere tanta locis.

The Stone was often regarded by the alchemists as a ' child of fire ', and Maier symbolises this idea in a plate entitled: ' As the Salamander lives by fire, so doth the Stone '. According to the epigram, ' the Stone does not reject the fierce burning of flames, for it was born in constant fire. The Salamander, being cold, quenches the heat and comes forth free; but the Stone is hot, and therefore heat, being like it, agrees with it.'

This summary of what is probably the most interesting of Maier's alchemical works may be aptly concluded by quoting some sage advice which he gives to the seeker after chemical knowledge: ' To him who is occupied with Chymistry, let Nature, Reason, Experiment, and Reading be his leader, his staff, his spectacles, and his lamp. Be Nature thy guide, and do thou follow closely after her gladly; thou dost wander unless she herself is thy travelling-companion. Let Reason help thee like a staff, let Experiment strengthen thine eyes, to see that which is situate afar off. Let Reading be the lamp shining in the darkness, that thou mayst foresee and avoid the heaps of facts and of words.' 35

The extracts which have been given from *Atalanta Fugiens* are sufficient to show that Maier considered the secrets of the Hermetic Art to lie concealed in the mythology of the ancients. The mere title of this work, with its suggestion of an analogy between the pursuit of fleet-footed Atalanta and that of the elusive Stone, affords an indication of Maier's idiosyncrasy. Curiously enough, this view of alchemy was developed with a still greater extravagance by Pernety, not long before the time when Lavoisier elaborated the revolutionary ideas which raised chemistry to the status of an exact science. Pernety's views may be gathered from a work of more than ordinary interest to the modern student of alchemy, which was published at Paris in 1758 under the title: *Dictionnaire Mytho-Hermétique, dans lequel on trouve les Allégories Fabuleuses des Poètes, les Métaphores, les Énigmes et les Termes barbares des Philosophes Hermétiques expliqués.* According to Pernety, the Greek fables were derived from ancient Egypt. Their author was Hermes, ' cet homme célèbre dont la mémoire sera éternellement en vénération ', who imparted the secrets of the ' Divine Art ' to chosen priests by means of enigmas, parables, allegories, fables, and hieroglyphics, all of which he invented for this particular purpose! It is thus in the materials and operations of the Divine or Hermetic Art, says Pernety,[36] that one should seek for the origins and real meanings of Egyptian and Greek mythology. It has been shown in the earlier part of the present work that the adepts were wont to compare their theories and operations with everything which they held to be of significance in the macrocosm: because of this, a definite interest attaches to expositions of the relationships which some of them imagined to exist between alchemy and mythology.

§ 6. ALCHEMY AND MUSIC

Smells are surer than sounds or sights
To make your heart-strings crack.

This dictum of a modern poet is amply borne out in the history of chemistry. At the very dawn of alchemy, smell leaped into first place in the sensual hierarchy of the sons of Hermes, and it has maintained its position ever since. Aqueous sulphuretted hydrogen, under the name of Holy Water, gave rise to the odour of sanctity in ancient Egypt (p. 14); and Chaucer, in the *Canones Yeomans Tale*, stresses the odoriferous character of the fourteenth-century alchemist: [37]

And evermore, wher ever that they gon,
Men may them knowe by smellyng of bremstoon;
For al the world thay stynken as a goat;
Their savour is so rammyssh and so hot,
That though a man fro them a myle be,
The savour wil infecte him, truste me.

The production of the Great Work, or Philosopher's Stone, was often supposed to be heralded by the advent of a powerful odour. Nicolas Flamel writes: ' Finally, I found that which I desired, which I also soone knew by the strong *sent* and *odour* thereof' (p. 65). Similarly, in Cremer's *Testament*, it is stated: ' When this happy event takes place, the whole house will be filled with a most wonderful sweet fragrance; then will be the day of the Nativity of this most blessed Preparation ' (p. 172).

Next to smell, in this alchemical hierarchy, came colour, the importance of which in the eyes of alchemists of all ages has been emphasised throughout this book. ' Red is last in work of *Alkimy* ', wrote Norton, alluding to the consummation of the Great Work; and Ripley described in the following words the spectacular succession of colours which marked the progress of the Work (p. 147):

Pale, and Black, wyth falce Citryne, unparfyt Whyte & Red,
Pekoks fethers in color gay, the Raynbow whych shall overgoe
The Spottyd Panther wyth the Lyon greene, the Crowys byll bloe
    as lede. . . .
And after all thys shall appere the blod Red invaryable.

That sound, also, should be taken into account by the alchemists is not surprising when it is recalled that many of the leading ideas of alchemy may be traced back to the Greeks (p. 11). Among such ideas which played a basic part in alchemy was the Pythagorean conception that the Cosmos had its origin and interpretation in Number. Later, Plato developed the view that Nature is based upon a mathematical plan, and that the ultimate realities must be sought in mathematics. His belief, stated in the *Timaeus*, that matter is an expression of mysterious harmonies, is also in keeping with the thought of the Pythagorean schools: for the Pythagoreans had discovered that definite numerical relationships exist between the notes of musical scales. Thus, the numerical theory of the Cosmos led the Greeks to attach particular importance to music.[38] Doubtless, in some form or other, music was almost coeval with speech; but it is to Greek music that mediaeval and modern forms must be traced.

Alchemy derived its mystical relationships of a numerical kind partly from the doctrines of the Pythagoreans and partly from those of cabbalism. According to the latter system, words could be represented by numbers. Thus, gold had the value 192, or $(1 \times 2 \times 3 \times 4)$ 8; moreover, the name of gold, the king of the mineral world, stood in a mystical relationship to the four-lettered Tetragrammaton, or Name of the Lord, the King of the spiritual world.[39] Numbers were sometimes used to designate the colours of the Great Work, the consummation of which was stated by Philalethes to be concealed in the numbers 448, 344, 256 and 224. Many references have already been made (pp. 9, 208) to the mystical significance of number in alchemy, and to the particular importance which was attached to the first four terms in the natural series of numbers. The summation of these numbers, in the tetractinal operation of Pythagoras, led to the fundamental number ten $(1 + 2 + 3 + 4)$, which played so important a part in the cabbalistic theory of the creation. The number ten, derived in a similar way (p. 208), was assigned also to the Philosopher's Stone.

Besides the four elements, the cabbalists recognised many other mystical groups of four, such as the four seasons, the four corners of the earth, the four winds, the four rivers of Paradise, the four spirits, and the four guardian angels. To this list, a later generation might be tempted to add the four valencies of the carbon atom and the four points of the tetrahedron. As Hoefer [39] remarks: 'mystical combinations based upon numbers, are, it will be said, nothing more than reveries of an ancient day. Well and good. But in our own time, when so much authority is attached to experiment, has a better explanation been devised for atomic combinations, based upon arithmetic and geometry?'

The Pythagoreans held likewise that the positions and movements of the heavenly bodies were determined by numerical laws, and that their harmonic motions produced a species of celestial music, known as 'the music (or harmony) of the spheres'. This conception was still held in the seventeenth century. Shakespeare gives

PLATE 59

(i) The Eagle, Lion, and Salamander.

From *Viridarium Chymicum*, Stolcius, 1624 (reprinted from *Philosophia Reformata*, Mylius, 1622). (See p. 267.)

(ii) The Approach of the Black Crow.

From the same. (See p. 267.)

PLATE 60

(i) Another Synthetic Emblem of the Great Work.

From *Viridarium Chymicum* Stolcius, 1624. (See p. 270.)

(ii) 'I am become as the Black Crow.'

From the same. (See p. 270.)

PLATE 61

Mortification, or the Death of the King.

From *Viridarium Chymicum*, Stolcius, 1624 (reprinted from *Philosophia Reformata*, Mylius, 1622). (See p. 271.)

PLATE 62

The Whole Work of Philosophy.

From *Viridarium Chymicum*, 1624.

eloquent expression to it in a celebrated passage in the *Merchant of Venice*: [40]

> Look how the floor of heaven
> Is thick inlaid with patines of bright gold:
> There's not the smallest orb which thou behold'st
> But in his motion like an angel sings,
> Still quiring to the young-eyed cherubins;
> Such harmony is in immortal souls;
> But whilst this muddy vesture of decay
> Doth grossly close it in, we cannot hear it.

Pythagorean ideas find repeated expression in alchemical literature (p. 209), and these ideas were closely associated with music. Moreover, from very early times music has held a prominent part in the ritual and ceremonial of religion, magic and necromancy. From many sides, therefore, alchemy was brought into intimate touch with music.

It should be noted also that music flourished among the Arabs; indeed, music, like alchemy, developed in Western Europe under Muslim influences. As Sarton [41] points out, many musical terms bear witness to the Arabic origin of European music, for example, the names of certain instruments, such as the lute, guitar, rebec, canon, eschaquiel or exaquir, etc.

There is even a record of a musical alchemist as far back as the tenth century, when a certain Abou-Nasr-Mohammed-Ibn-Tarkan, otherwise known as Farabi or al-Farabi, is said to have flourished in Asia Minor and Syria. He is represented as an author of works on alchemy and music, and also as an accomplished performer on the lute. ' At the request of the Sultan [of Syria],' writes Waite,[42] ' he produced a piece of his own composing, sung it, and accompanied it with great force and spirit to the delight of all his hearers. The air was so sprightly that even the gravest philosopher could not resist dancing, but by another tune he as easily melted them to tears, and then by a soft unobtrusive melody he lulled the whole company to sleep.'

PRELUDE TO CHEMISTRY

Sometimes alchemical designs contain representations of allegorical and mythological figures bearing musical instruments. The medallion at the head of the title-page of the *Musaeum Hermeticum* of 1625 and 1678 (p. 167) provides an example (Plate 30). The central figure is playing upon a lyre; two of the others have trumpets, one a harp, and one a viol; and a lute lies upon the ground. Minerva, shown on the left of the medallion, was reputed to be the inventor of musical instruments in general and of wind instruments in particular; Hermes, the alleged founder of the Hermetic Art, shown on the right, was also regarded as a patron of music. The same volume contains an illustration of seven figures, representing the seven metals, seated in a cave (Plate 31). The central figure is again playing upon a lyre. This emblem represents a blending of astral (p. 197) and musical influences in the production of metals in the bowels of the earth. A similar drawing (Plate 57 (ii)), given by Mylius and Stolcius, depicts Sol playing upon a harp, and possibly represents 'the Chymic choir', in which, according to Stolcius (p. 258), Miriam the sister of Moses rejoiced and triumphed.

That musical influences were held to be essential in the laboratory operations of alchemy, as well as in Nature, is clear from Norton's *Ordinall* of 1477, in which the following significant passage occurs: [43]

> Joyne them together also *Arithmetically*,
> By suttill Numbers proportionally, . . .
> Joyne your Elements *Musically*,
> For two causes, one is for Melody:
> Which there accords will make to your mind,
> The trewe effect when that ye shall finde.
> And also for like as *Diapason*,
> With *Diapente* and with *Diatesseron*,
> With *ypate ypaton*, and *Lecanos muse*,
> With other accords which in Musick be,
> With their proporcions causen Harmony,
> Much like proportions be in *Alkimy*,
> As for the great Numbers Actuall:

A MUSICAL ALCHEMIST

But for the secreate Numbers Intellectuall;
Ye must seeche them as I said before,
Out of *Raymond* and out of *Bacons* lore.

Norton here attaches a truly Pythagorean emphasis to Number, Harmony, and Music. The terms diapason, diatesseron and diapente, indicate the musical intervals of an octave, a fourth, and a fifth, respectively; ypate (ὑπάτη), ypaton (ὑπατῶν), and Lecanos muse (λιχανός, forefinger; *mese*, or μέση), are various notes of the tetrachord. By ' accords ', Norton does not mean chords in the modern sense: he refers rather to notes having sound-values (expressed by relative lengths of strings) which fit together into a scale on intelligible principles, and so ' with their proporcions causen Harmony '.

As we have seen (p. 179), Norton was a working alchemist with a strong sense of practical values; but the extent to which he, or others, endeavoured to influence their laboratory operations by means of music is not clear. However, in view of the alchemical belief in the beneficent influence of music, it is likely that the processes of the Great Work were sometimes performed to the accompaniment of musical chants or incantations. To the religious mystics among the alchemists, in particular, these processes partook of the nature of a religious ritual, and it would be natural for them to introduce music from one of these closely related activities to the other.

The exercise of combined astrological and musical influences upon the processes of the Great Work are suggested in an illustration from *Splendor Solis* (1582), reproduced in the Frontispiece.

A more suggestive illustration (Plate 52) occurs in Heinrich Khunrath's *Amphitheatrum Sapientiae Æternae*, of 1595 and 1609. This depicts a laboratory which is also an oratory. Khunrath, the celebrated Hermetic mystic (p. 81), is shown in the act of praying at the oratory. Behind him, among the stills and furnaces in the laboratory, may be seen a number of musical instruments, including two lutes, a viol, and a harp; the table which bears them is ornamented with a Latin motto stating that ' sacred music disperses sadness and malignant spirits '.   It is also significant that the praying figure is gazing at a representation of the pentagram, the badge of the Pythagorean Brotherhood.[44]

Altogether, it is likely that the idea of the operation of a musical influence in alchemy, and in Nature generally, received more credence from the mystical alchemists and philosophers than from their severely practical brethren, the so - called ' puffers ' (p. 2).   Thus, the English mystic, Thomas Fludd (1574–1637), who, as ' a notable Christian theosophist '[45] had much in common with Khunrath, wrote at great length on the harmonies of the world and the music of the spheres in a formidable work published by J. T. de Bry at Oppenheim in 1617 under the title, *Utriusque Cosmi Maioris scilicet et Minoris Metaphysica, Physica atque Technica Historia*.   Here, in involved and obscure language, he discourses upon the world monochord, world-music, the determination of the harmonies from which the world harmony is produced, the nature of world discords, and so forth.   The compositions of the Aristotelian elements are supposed to be capable of numerical expression: for instance, coldness (p. 10) is present in earth and water in the proportion of 4 to 3.   Similarly, ' if we can ascend into the first step of the ethereal region, we shall find the sphere of Luna has reference to the earth in the proportion of one and a half '.   Fludd likens ratios of this kind to musical intervals, such as diapason, diatesseron, and diapente. Moreover, he draws the planets and zodiacal constellations into the scheme: so that, for example, ' Leo receives Sol as his master, whence Leo is, as it were, of one sound with the nature of Sol; so also Cancer with Luna, etc.' One of several synthetical diagrams embodying schemes of this kind is reproduced in Plate 53: in it, Fludd depicts the hand of the Master Musician attuning the harmonies of the Aristotelian elements and the seven heavenly bodies and their corresponding metals into ' one grand, sweet song '.

Probably the last attempt, and certainly the most definite and impressive, to link alchemy with music was made by Michael Maier in *Atalanta Fugiens*. Maier's idea to set his Latin epigrams to music may possibly have originated during his translation of Norton's *Ordinall* into Latin verse: the translation was published in his *Tripus Aureus* in the same year which saw the appearance of *Atalanta Fugiens*. He may also have derived some inspiration from Fludd's massive work, which was issued by Lucas Jennis at Frankfurt in 1617, a year before the production of *Atalanta Fugiens* by the same enterprising firm. Both Maier and Fludd were exponents of Rosicrucianism, but in spite of their common interests there is no direct evidence that they met during Maier's visit to England (p. 229).

In examining Maier's music in *Atalanta Fugiens*, modern chemists who may wish to sing these alchemical lays should not overlook the notice to music-loving readers which is inserted at the end of the book. ' I had hoped hitherto ', writes Maier, ' to play Prometheus rather than Epimetheus [*i.e.* to be wise before and not after the event], and so I wrote and revised with the greatest care everything which concerns the fugues. However, good counsels sometimes fail of their effects, and whether it was in my absence or not, many notes of these fugues have somehow got transposed and not printed correctly as was to be wished. Therefore I am forced to add this note of errata in the foregoing songs especially. Anyone skilled in music will easily perceive, even if I do not draw his attention to it, that each fugue should be wholly like the rest in values, the rise and fall of each note, save that one sound follows another in different keys. Therefore he will be able to correct for himself any printer's error, whenever he meets with anything set down otherwise than rightly. Meantime, by way of example, I will put down here a few mistakes, but not in the music, which I have noted in passing.' The educated man of Maier's day was evidently assumed to have a knowledge of music as well as of alchemy.

To what extent Maier's alchemical canons were sung in the laboratory, or elsewhere, is unknown, as no records of their performance have been discovered. Some examples of these canons in their original and transcribed forms, together with a descriptive note, are given in the Appendix to this book.[46]

Music at the Close of the Great Work.
From *Elementa Chemiae*, Barchusen,
1718.

CHAPTER VII

THE GARDENS OF HERMES

So the Gardener had them into the Vineyards, and bid them refresh themselves with Dainties. He also shewed them there the King's walks, and the Arbors where he delighted to be. And here they tarried.

JOHN BUNYAN

§ 1. BOHEMIA'S STUDENT-POET OF ALCHEMY

 YOUNG student of medicine, who had set forth from his native Bohemia on foreign travel in order to widen his knowledge, found himself in the summer of 1623 in far-off Oxford, where he wrote the dedication of one of the most remarkable of all alchemical works. His name was Daniel Stolck, Stolcius, or Stolcio, of Stolcenberg. A humble disciple of the great Michael Maier, Count Palatine and learned Doctor, this modest student is an important although forgotten figure in the history of the Frankfurt Emblems. Originally a graduate of the University of Prague,[1] Stolcius matriculated at the University of Marburg in 1621, and as an ardent admirer of alchemy and alchemical literature he was naturally drawn to Frankfurt-on-Main in the course of his wanderings. In this renowned publishing centre he called upon the firm of Lucas Jennis, and here he was shown a number of alchemical engravings which aroused in him a desire to publish a work of his own which would bring pleasure to himself in his exile and also to other lovers of the 'Divine Art'. With the help of Lucas Jennis he assembled a large number of copper-plates which had

<CENTER>255</CENTER>

been used in the course of the past few years to illustrate various alchemical books. Jennis promised to publish the collection, provided that Stolcius, who was a laureated poet, would compose an apposite Latin epigrammatic verse for each figure. Although this entailed much reading and comparison of alchemical writings, Stolcius undertook the task. The outcome of his work was the *Viridarium Chymicum*,[2] or 'Pleasure Garden of Chymistry', published by Lucas Jennis at Frankfurt in 1624.

This fascinating little volume, of which the title-page is reproduced in Plate 54, was evidently regarded as an alchemist's *vade mecum*; oblong in shape, and measuring about 6 by $3\frac{3}{4}$ inches, it slips seductively into the pocket. In its original state, it is one of the rarest of alchemical works. A German edition was issued by Jennis in 1624, and this was reprinted in 1925 at Leipzig, with somewhat indifferent reproductions of the copper-engravings which are so beautifully clear in the original edition. The author describes it as 'The Pleasure Garden of Chymistry: accompanied by copper-plates, and adorned with poetical pictures; so as not only to please the eye and amuse the mind, but to arouse a deeper contemplation of Nature; convenient also, from its oblong shape, for a friend's album'. Stolcius expressed his gratitude to his publisher by incorporating his name in an introductory anagram, acrostic, and chronodistichon for the year 1624. The anagram is *Lucas Jennis—Lucina Senis* (the ancients' Lucina, or birth-goddess). A Latin verse explains the allusion: 'Of old, Lucina lent aid at births, that the babe might arise under favourable auspices. You search out the sayings of old Hermes, and in word and deed help them to see the light.'

The dedication of the work is dated Oxford, England, 6/16 July 1623. In his preface Stolcius refers to the troubles of his country in the unhappy conflict which in due course developed into the Thirty Years' War (1618–1648); and the succeeding verse in the language of the

THE GARDENS OF HERMES

Czechs contains a hint of home-sickness. ' I wished ', he writes, ' to provide myself with a repository of things I like, to refresh my eyes with artistic representation and to delight my mind with its hidden meaning; especially in this my journey abroad, undertaken in the cause of Medicine, in which I hear with grief of the strange and pitiable disasters of my country, and am, to my extreme sorrow, very often interrupted by these tumults of war that are scattered everywhere. . . .

' Therefore, kind Reader, use and take pleasure in these as seems best to thee, and take a pleasant stroll in my garden. Thank the cherished memory of the very famous and learned Herr Michael Maier, the most cele-brated Doctor of Physic and Medicine, for part of the illustrations; Master John Daniel Mylius, that industrious Chymist, for the rest.'

The verse in the Czechish tongue runs thus: ' Wonder not, dear Reader, that thou art given naught here. Thou wilt surely receive rarer gifts when I return, I promise thee. Therefore do thou wish Godspeed and a quick home-coming to thy faithful and loyal fellow-countryman.'

The ' Pleasure Garden' consists of a collection of 107 copper-engravings, each of which is provided, on the opposite (left-hand) page, with a Latin epigram in the form of a six-line verse provided with a short title. The back of each leaf is blank. The first fourteen of the figures, beginning with the Twelve Keys of Basil Valen-tine, are printed from plates used in the *Tripus Aureus*.[3] The next figure,[4] entitled ' Mercury, Lord of all Worldly Things ', represents a fantastic parliament of beasts and other things, of which details are given in Michael Maier's *Lusus Serius* (' Serious Pastime '), published by Lucas Jennis at Frankfurt in 1616. ' It is difficult to see that a *jeu d'esprit* of this kind has any title to existence, but it found appeal in two quarters, being reprinted at Oppen-heim in 1619 and translated into English.' [5]

Next in order, in the *Viridarium*, follow the emblems [6] of the Twelve Chosen Heroes of Maier's ' Contributions

of Twelve Nations to the Golden Table ' (p. 223), printed
from the plates used in that work. The sole chosen
Heroine, Miriam the Jewess (Plate 2 (i)),[7] is described in
the epigram as follows: ' Of the race of Palestine, sister
of Moses, behold Miriam equally rejoices and triumphs
in the Chymic choir. She knew the hidden secrets of
the great stone. She has made us also learned, sage that
she is, with her words. Smoke loves smoke, and is loved
by it in return: but the white herb of the lofty mountain
captivates both.' The white herb, which may be Lunary
(p. 97), represents the white stage of the Great Work
(p. 146); white and red smokes signify mercury and
sulphur of the philosophers. Stolcius' epigrams are
clearly modelled upon those of *Atalanta Fugiens*; as an
example of their general form, the Latin original of the
epigram just quoted is reproduced below:

MARIA HEBRÆA

Gente Palæstina Moysis Soror, ecce Maria
In Chymico pariter gaudet ovatq; choro.
Abdita cognovit lapidis mysteria magni.
Erudijt dictis nos quoque docta suis.
Fumus amat fumum, rursusq; adamatur ab illo:
Alba sed herba alti montis utrumque capit.

Avicenna the Arabian (Plate 2 (ii))[8] ' made known
to the world the secret of the Magisterium and inserted in his writings fair emblems ', says another epigram, which then proceeds, ' Join the toad of earth to
the flying eagle and you will see in our art the Magisterium '. This seems to mean that the seeker after the
Philosopher's Stone should combine the fixed and volatile
principles—sophic sulphur and mercury. ' The vulture
flying in the ayre, and the Toade going upon the Earth,
is our Maistry ', writes Artephius [9] in similar vein.
Roger Bacon the Englishman (Plate 41 (ii)) [10] is
introduced in the following words: ' Lo, Bacon the monk,
Professor at Oxford, adorned by his labours the English

countryside. He published the greatest marvels of the universe which in the process of Nature any can have. When you shall make equal the weights of the elements, you will behold with your eyes welcome gifts.'

The five illustrations [11] succeeding those of the Twelve Chosen Heroes in Stolcius' *Viridarium Chymicum* are reprinted from the actual plates used in Michael Maier's *Septimana Philosophica* (' Philosopher's Week '), published by Lucas Jennis at Frankfurt in 1620. This work, described on the title-page as ' Golden Enigmas on the whole system of Nature ', takes the form of imaginary conversations between King Solomon of Israel, King Hiram of Tyre, and the Queen of Sheba. The five illustrations are entitled Meteors, Minerals, Vegetables, Animals, and Man.

In his epigram dealing with Maier's illustration of the vegetable kingdom, Stolcius outlines his ideal alchemical garden. ' As a garden grows green with excellent herbs and plants,' he says, ' so this our garden presents a many-coloured spectacle. On this side is the iris, the vine, lunary and moly, corn-harvests, and thy flower, oh blushing rose, the fruits of the Hesperides, the mulberry and runaway Daphne, also the golden bough, the myrtle, olive and saffron.' Each of these items has a definite alchemical significance. The iris is reminiscent of the rainbow colours of the Great Work (p. 146); the vine is a symbol of fruitfulness; lunary is the potent signature-herb of Luna (p. 98); moly is the magic herb with black root and white flower given by Hermes to Ulysses as a counter-charm against the spells of Circe—' hard for mortal men to dig, howbeit with the gods all things are possible ' [12] ; the corn-harvest, particularly wheat, represents cibation (p. 139), and also the vital principle (p. 95); the rose is Ben Jonson's ' flower of the sun, the perfect ruby, which he calls elixir '; the golden apples of the Hesperides and the golden bough are symbolical of the gold-making powers of the Stone; the mulberry ministers to the mysterious transmuting powers of the silk-worm; the myrtle is a symbol of immortality; the olive is sacred to Minerva, the goddess of wisdom; saffron typifies the power of dyeing or tingeing, which was associated with the Philosopher's Stone (p. 206). Runaway Daphne is the laurel, because Daphne ran away from Apollo (Phoebus) and was turned into a laurel-bush: in alchemy, Phoebus and Daphne sometimes represent the fixed and volatile principles, sometimes the masculine and feminine, as exemplified in an old English poem contained in Elias Ashmole's collection (p. 163). Truly, Stolcius' alchemical garden is well stocked!

§ 2. THE EMBLEMS OF MYLIUS

Of the remaining seventy-five illustrations contained in the *Viridarium*, sixty-one [13] are reprinted from the actual plates used in a work by Johann Daniel Mylius, ' student of Theology and Medicine ', of the Wetterau in Hesse. This is a rare volume of more than 700 pages, written in Latin, and published by Lucas Jennis at Frankfurt in 1622, under the title *Philosophia Reformata* (' Philosophy Reformed ').[14] It consists of Book I, an Epilogue to Book I, and Book II. According to the expansive title-page, Book I deals with the generation of the metals in the bowels of the earth, expounds the twelve stages of the wise philosophers, and, in general, reviews the theory and practice of the Divine Art. Book II contains the authorities of the philosophers, that is, quotes opinions of prominent writers in support of Mylius' doctrines. The preface ends, in true alchemical style, with a chronogram of 1620: ' Frankfurt-on-Main, from my study, in the great month of August, in that ominous year which *weeps so that the skies fall (Flet, Vt MVnDVs CorrVat)* '.

In spite of its pretentious title and an introductory flourish of trumpets, *Philosophia Reformata* contains no new ideas or observations of interest. The first book opens with the statement that ' all metals are generated of the vapour of sulphur and mercury. For when the fatness of the earth, warmed by water, has found a substance somewhat rounded, it has solidified the said substance, by the action of its natural virtue and by the influence of the heavenly bodies and the force of the heavens, according to the purity or impurity of both, into gold, silver, copper, tin, iron, and lead, those most splendid bodies, in the veins of the earth.'

The chief interest of this portentous publication lies in its illustrations; in this respect, as in others, the alchemical productions of Mylius are closely similar to those of his contemporary, Maier. In addition to the engraved title-pages of its two books, *Philosophia Reformata* contains sixty delicately executed drawings of uniform size,[15] arranged in blocks of four (Plate 55). These sixty illustrations[16] are reproduced in Stolcius' *Viridarium Chymicum*, each with an original epigram of the usual form.

The engravings are all contained in the first book of Mylius, and the first sixteen of them brighten up the otherwise dull disquisition on the twelve stages of the Great Work.

The first[17] engraving (Plate 55) depicts the Stone in the guise of an infant, whose nurse, according to the fourth precept of Hermes (p. 54), is the earth; the feet of the nurse are immersed in the second element, water; the birds signify air, and the salamander stands for fire.

The second[18] picture shows, in the words of Stolcius, ' four sisters of equal rank ' balancing themselves upon four globes marked with the symbols of earth, water, air and fire. ' The first bids thee dissolve the body first selected, the second to wash properly thy matter; the third advises to join the separated portions; the fourth teaches thee to fix thy Stone in fire.'

The three-headed serpent in the third[19] picture signifies that the Stone is composed of spirit, soul, and body, proceeding from a common source; it is single in essence but triple in form, ' and so ', adds Stolcius, ' the offspring of Sun and Moon gets hold of three natures '. The four Vases containing the crow, peacock, swan, and king symbolise various stages in the Great Work of preparing the Philosopher's Stone (p. 146).

Stolcius' epigram upon the fourth [20] drawing runs thus: ' With his skilled arrows the god of Delos [Apollo] lays stiff Python low, that he may lead his life in fire. But if anyone would know who that dragon is, lo, the ancients say it is their sulphur. Yet, if you desire to know whence he of Delos gets his bow and arrows, the couchant lion will give thee knowledge.' Here is another of the many representations of the sulphur-mercury theory, the two principles being symbolised by the lion and dragon, respectively. The epigram refers to Apollo's first exploit in killing Python, the dragon of Delphoi, either because it was molesting his mother, Leto (or Latona, p. 161), or because it opposed him when he went to set up his shrine there.

Mylius then proceeds to discuss each of the twelve operations of the Great Work, which he arranges in the following order (cf. p. 136): calcination, solution, separation, conjunction, putrefaction, congelation, cibation, sublimation, fermentation, exaltation, multiplication, and projection. Each process, except projection, is represented by means of an emblematic drawing.[21]

The drawing illustrating *calcination* is a slightly modified reproduction of the illustration of the Twelfth Key of Basilius (Plate 38 (ii)): it shows fire, the calcining agent, and the victory of the fixed principle (the lion) over the volatile (the serpent). Then, as Stolcius explains, the volatile spirit, having been fixed and joined to earth, brings forth the flowers of the noble metals, silver and gold. *Solution* is ' the reduction to a certain liquid substance of the hard matter, dry earth. Hence the philosopher says, *Until all be made water, perform no operation.*' The corresponding illustration is modelled on the Fifth Key of Basilius.

The third stage, which is symbolised in a manner

suggestive of the Second Key of Basilius, ' is called our *separation*, which is the purifying of watery vapour or liquorous matter from its dregs'. *Conjunction*, defined as ' a combination of qualities, an equalisation of principles, a putting in order of contrariety ', is symbolised by a representation of a marriage ceremony between the king and queen, with Sol and Luna in the sky, and a rainstorm, signifying condensation and fertility, in the background. In the foreground are represented ' a fiery man with two heads ', and Neptune, lord of the sea, about to prepare a water-bath. All these features occur also in the Sixth Key of Basilius.

The seed of the metals, resulting from conjunction, is next submitted to *putrefaction* (p. 138). According to Mylius, this process ' is carried on for six and forty nights in the shade of purgatory, and is a necessary putrefaction, because it is corruption and a making ready of regeneration '. The accompanying symbol is suggestive of the Fourth Key of Basilius: it depicts a skeleton standing upon a black globe, with a black crow perched upon an uplifted hand. Putrefaction is often denoted in alchemy by blackness. *Congelation* is defined by Mylius as ' gentle induration, hiding of moisture, fixation of spirit, renovation of the homogeneous, a suitable adaptation to fire of those which fly from fire '.

Cibation, represented by a suckling infant (p. 139), is said by Mylius to result in ' an augmentation of whiteness and redness, goodness and quantity, and a complete laying bare of the variety of colours '. The portrayal of *sublimation* (Plate 56 (i)) contains the four main features of the First Key of Basilius, grouped similarly: namely, the King, Queen, wolf and crucible, and Saturn with a scythe; the King, however, carries the sign of the phoenix and the Queen that of the swan. Mylius remarks that ' the modes of our sublimation and exaltation do not differ essentially '.

Fermentation, according to our author, ' is the incorporation of the living thing, and it is two-fold, namely, white and red ': the illustrative design is based closely upon the Eighth Key of Basilius (Plate 37 (ii)). The symbol of *exaltation* follows a design of Libavius (Fig. 16), and shows a series of steps guarded by lions and leading up to the throned King and Queen, with the tree of the metals between them.

Mylius gives the following ' reason ' why ' our medicine ' should be capable of indefinite *multiplication*: ' the medicine itself is like fire in sticks, or like musk in good perfumes; therefore it increases more, the more it is crushed with handling '. The illustrative engraving (Plate 56 (ii)), inspired by the Eleventh Key of Basilius, shows the Queen riding upon a lion with numerous whelps; in her hand she carries a Vase of Hermes enclosing a pelican, also provided with a numerous progeny. Finally, the twelfth operation of *projection* is likened to dyeing with saffron, which, if applied dry, ' dyes but little; but if being dissolved it is joined to a little liquid, even a little thereof added to much would dye endlessly. Thus therefore you shall make the projection. First multiply ten by ten, and they will be a hundred; then a hundred by a thousand, and they will be ten thousand [*sic*], and so on by the usual rules to infinity.'

In the course of his remarks upon the foregoing operations, Mylius lays down four degrees of heat: ' the first, slow and mild, as of the flesh or the embryo; the second, moderate and temperate, as of the sun in June; the third, great and strong, as of a calcining fire; the fourth, burning and vehement, as of fusion. Each of these is twice as great as the preceding degree.' Stolcius' epigram upon the relevant illustration (Plate 23 (i))[22] states that ' Phoebus [the sun] runs through the zodiac in a year, and recreates all seeds with his rays. Hence learn the four degrees of our toil . . . they are the Ram, the Crab, the Scales, and Capricorn [the signs of the four seasons, the two equinoxes and two solstices].'

The remaining thirty-two engravings[23] in Book I of *Philosophia Reformata* illustrate some of the foregoing and

PLATE 63

Four English Alchemical Philosophers.

From *Bibliotheca Chemica Curiosa*, Manget, 1702 (reproduced from *Hortulus Hermeticus Flosculis Philosophorum*, Stolcius, 1627).

other related themes. Thus, the pelican nourishing its young by thrusting its red bill into its own breast is used as a symbol of revivification: 'Thy chicks, Pelican!' exclaims Stolcius,[24] referring in his epigram to the old superstition so freely used in alchemy, 'Thy chicks, thy dearest offspring, thou dost nourish with the blood which trickles from thine own bosom: whence fresh virtue returns to their hearts, and life-giving vigour warms their tender limbs'. Another engraving [25] depicts a king arising from the tomb, as a symbol of resurrection (p. 202). According to Stolcius, the king has been defeated by a servant who 'dissolves his limbs by means of his own seed:' this veiled reference to the amalgamation of gold and mercury is closely similar to the imagery of the *New Pearl of Great Price* (p. 58).

An engraving [26] showing a spirited encounter between a maned and a winged lion (Plate 57 (i)) is reminiscent of several of the illustrations in *Atalanta Fugiens*. It is entitled 'The Two Sulphurs', and the 'interpretation' given by Stolcius runs: 'Behold, twain lions come together with their feet joined, and enter upon a firm pact of friendship. Thou who seekest a ferment, join the two sulphurs to be able to multiply thy mass. Let one be constant, let the other fly upwards, but being joined let them stay and remain with agreeing tread.' The winged lion represents mercury, the volatile principle; the other symbolises sulphur. A favourable conjunction of the two gives rise to metals and even to the Philosopher's Stone (p. 133).

The seven metals themselves are symbolised in an engraving (Plate 57 (ii))[27] in which, as Stolcius says, 'thou seest laid open the bowels of the vast earth'. The drawing gives expression to the belief that the generation of metals in the earth is controlled by the stars, as well as by the sun, moon, and planets with which the metals are immediately joined. The metals are represented, in order, by figures of the seven Olympian deities, Jupiter, Mars, Saturn, Mercury, Diana (Luna), Apollo (Sol), and

PRELUDE TO CHEMISTRY

Venus (p. 88): these are accompanied by zodiacal
symbols of the Archer, Ram, Goat, Fish, Crab, Lion, and
Bull. Saturn (Kronos) biting an infant, which he carries
in his arms, is suggestive of the mythological story (p. 243)
and of the use of the term ' infants' blood ' for the mineral
spirit of the metals by Flamel (p. 63). The harp borne
by Sol probably symbolises the music of the spheres
(p. 248). The four circles signify the four elements.
' Out of doubt ', says the epigram, ' the earth itself
possesses its own planets, to which the elements furnish
their powers. If thou dost doubt what they are, with
watchful mind contemplate the metals, so shall the sky
be known unto thee.'

The celebrated Bath or Fountain of the Philosophers
is depicted in an illustration [28] bearing a fundamental likeness to an earlier drawing given by Libavius (Fig. 16,
p. 220). ' Take our water with which no limbs are wet
[sophic mercury],' says Stolcius; ' make Sun and Moon
bathe in it together. When thou hast accomplished this,
let a Breath be combined with them; thereafter shall
thine eyes behold two lilies. Hence what tree thou wilt
shall arrange its own fruits on a tree, and therefrom thou
mayest pluck thee the fruit.' At the sides are shown
the sun-tree and moon-tree, symbolising the Great
and Little Work (p. 131). The recurring idea of the
Bath of the Philosophers is connected with the mythological story of the son whom ' gentle Venus bore to
Hermes ' (p. 239). The incorporation of this beautiful
youth with the nymph Salmacis, while bathing in a fountain, gave rise to the single being Hermaphroditus. The
property acquired by these mythological waters, of imparting a bisexual nature to those who bathed in them,
was supposed to be shared by sophic mercury, and the
stage of the Great Work attained after the alchemical
process of conjunction was sometimes called Hermaphroditus, Rebis, or Androgyne. Flamel evidently alludes
to the Bath of the Philosophers when he describes the
great vessel containing infants' blood, wherein Sol and

Luna came to bathe themselves (p. 62): here, the fluid was the mineral spirit of the metals.

Some of the drawings occurring in Book I of Mylius' *Philosophia Reformata* and reproduced in Stolcius' *Viridarium Chymicum* are reminiscent of engravings in *Atalanta Fugiens*, the *Book of Lambspring*, the *Twelve Keys* of Basil Valentine, and other works. The divergencies are exemplified in the two renderings of the washing of clothes, in *Atalanta Fugiens* and *Philosophia Reformata*, which have already been given (Plates 45 and 23 (ii)), with their distinctive interpretations. Other drawings occur twice in Mylius' own collection, in slightly different forms.

An Epilogue to Book I of *Philosophia Reformata* [29] contains a print of an engraving showing two men conversing beneath the tree of universal matter (Plate 58). This was too large for inclusion in the collection of Stolcius; but the twelve other standard engravings of the Epilogue were duly reprinted from the same plates in *Viridarium Chymicum*. They depict the philosopher, the nymph of the sea, our sulphur, the philosopher's egg, generation, conjunction, mortification (Plate 61), putrefaction, albification, rubification, dream or vision, and symbol of our virgin.

All Stolcius' emblems from I to XCIII have thus been traced back to works published by Lucas Jennis at Frankfurt, between the years 1618 and 1622, and written by Maier or Mylius.

§ 3. 'AZOTH': A RIDDLE OF THE PHILOSOPHERS

It is a striking feature of Stolcius' collection that the next thirteen engravings,[30] although forming a distinct series of designs, are clearly intended to illustrate the same subjects as the thirteen preceding engravings taken from Mylius' Epilogue.[31]   Thus, Plates 59 (i) and 59 (ii), from the Epilogue, correspond respectively to the symbol on p. 211 and Plate 60 (ii), from the second series; the resemblance between Plate 58, from the Epilogue, and Plate 62, from the second series, is also evident.

Stolcius' second series of these designs are close copies of the drawings which appeared in an earlier tract, published at Frankfurt in 1613 by Johann Bringer under the title *Azoth, sive Aureliae Occultae Philosophorum* ('Azoth, or the Hidden Gold-making Secrets of the Philosophers').[32] Mylius' Epilogue consists of a repetition of *Azoth*,[33] with the addition of a preface couched in allegorical language: ' It came to pass once upon a time, that having spent many years of my life in sailing from the North to the South Pole, by the special Providence of God I was cast upon the coast of a certain great sea. . . . Now while I stood on the shore, watching the mermaids swimming this way and that with the nymphs, being moreover wearied with my previous toils and heavy with various cogitations, I was overcome by the great roaring of the water. And falling sweetly asleep, I was met in my slumbers by a wonderful vision. I beheld, coming forth from our sea, old Neptune with his reverend hoary locks, trident in hand, who greeted me friendly and then led me to a most pleasant island. This most lovely isle was situate towards the south, and well supplied with all things necessary for man and suited to delight him. . . .' Later, Saturn appeared upon the scene, and the revelations which were made in this ' wonderful vision ' are set down in the ensuing pages of the Epilogue.[34]

The first part of Mylius' Epilogue, as also of *Azoth*, takes the form of a dialogue between an adept called Saturnus, Senior, or Le Vieillard, and a neophyte named Philosophus, or Adolphus. Senior is here definitely identified with Saturn (pp. 156, 199). It is interesting, and possibly significant, that during the conversation Adolphus discloses that he comes from Hesse, which was the homeland of both Mylius and Johann Thölde, the supposed original of Basilius (p. 184).

The two characters are shown conversing beneath the tree of universal matter, whose fruits are the symbols of

the seven metals and heavenly bodies.   Between them,
in *Azoth* only (Fig. 17), is a triangle carrying symbols
of the four elements, sulphur, and mercury, and the
branches of the tree contain the symbol of fire.   The more

FIG. 17.—The Tree of Universal Matter, with Senior and
Adolphus.

From the title-page of *Occulta Philosophia von den verborgenen
Philosophischen Geheimnussen*, etc., Frankfurt, 1613.

elaborate illustration of Mylius (Plate 58) contains seven
inset symbols representing putrefaction, distillation, sublimation, the white (Little) Work, peacock, and red colours
of the Great Work, culminating in the birth of the Philosopher's Stone.   This drawing also depicts the fiery
Dragon issuing from his cave, the King (Sol) enthroned

upon a lion, and the Queen (Luna) seated upon a dolphin: the Queen here assumes the common guise of Diana carrying a bow.[35]

Some of the drawings occurring in the text of *Azoth*, as reproduced in Stolcius' *Viridarium*, may now be considered. The first[36] depicts Atlas supporting the universe upon his shoulders, and includes the celebrated ' vitriol acrostic ' (p. 155), which figures also in the next engraving (Plate 18 (ii)).[37]

Another emblem[38] shows the mermaid-like Aphrodite (Venus), goddess of the sea, with a golden crown and twin tails. Of this favoured symbol of the alchemists, the epigram in the *Viridarium* states: ' I am a goddess exceeding fair, born from the depths of the sea, which in its course washes and surrounds all the dry land. Let my breasts pour forth to thee twin streams of blood and milk [*lac virginis*], which thou canst well know. These two combined leave to be wrought upon by a gentle fire, then will the Moon and Apollo [the Sun] answer thy prayers.'

A further emblem (Plate 60 (i))[39] provides an interesting variant of the synthetic expression of the Great Work which has already been brought forward (Fig. 13). It shows the square and the triangle with the magic numbers 4 and 3, the dragon, and the philosophic egg. The two intersecting circles have been replaced, however, by the hermaphroditic figure of the ' Rebis ', or ' Two-Thing ', holding in its hands the geometrical symbols of set-square and compasses, which are required for constructing the square, triangle, and circle. The wings signify the volatile mercury, and the seven metals are shown by the customary signs; exceptionally, the symbol of Saturn is not darkened in this particular reproduction of the emblem.

An emblem[40] which has been widely reproduced in modern works is shown in Plate 60 (ii). ' Je suis vieil debile & malade ', begins the explanation in the French version of *Azoth*:[34] ' I am old, debilitated, and ill, my surname is Dragon : because of this, I have been shut up in a tomb so that I may attain the Royal Crown and enrich my family . . . my soul and spirit [the two winged figures] leave me, I am become as the Black Crow . . . in my body are found Sulphur, Salt, and Mercury.' In the *Viridarium*, this emblem is entitled ' The First Process '. It refers probably to the preparation of sophic mercury from ordinary mercury, or to the isolation of the seed of gold from base or earthy matter : through either of these processes the ' family ' of metals would be ' enriched ' by transmutation into gold. 'When the fortieth dawn returns I re-make the shining sceptre of my race ', states the epigram. A later emblem [41] in the series depicts the death of the King. Here, the fundamental idea is the same as that of the eighth Key of Basilius (p. 202) : death, followed by revivification, as processes in the Great Work. The epigram to a variant of this emblem (Plate 61), which is reproduced in the *Viridarium* [42] from the Epilogue to Book I of *Philosophia Reformata* (p. 267), is headed ' Mortification ', and runs as follows : ' Now the blazing glory of the King gives birth to envy ; and a band of ten rustic youths slay him. All things are in confusion, Sun and Moon in darkness reveal many signs of their sadness. There appears the rainbow painted in varied hues, which brings to the people glad tidings of peace.' There is a reference here to the black colour of the Great Work, followed by the rainbow colours of the peacock's tail (p. 146).

Finally, attention may be drawn to one of the most remarkable and complete of all alchemical designs (Plate 62).[43] Besides including all the usual geometrical symbols of the square, triangle, and circle (p. 239), it displays the ' vitriol acrostic '. The heavenly bodies and metals are denoted both by numbers and symbols, and the darkened symbol of Saturn points to the cubic Stone, which in turn is surrounded by five stars suggestive of the five planets. King Sol (Apollo), seated upon a Lion, is associated with earth and fire ; and Queen Luna (Diana),

riding upon a dolphin, with water and air.   The Dragon
with flaming jaws, which issues from a cave beneath the
King, probably signifies sophic mercury, in accordance
with the alchemical legend that the material of the Stone
is a black ball existing in hot subterranean caverns.   The
angles of the large triangle contain representations of the
assimilation of the Sun to the soul, the Moon to the spirit,
and the cubic Stone to the body.   There are also seven
symbols depicting putrefaction, distillation, and other
processes in the Great Work (p. 269).   The *Viridarium*
terms this emblem ' The Whole Work of Philosophy ',
and adds an interpretation in these words : ' Those which
aforetime were shut up in many shapes you now see all
included in one round.   The beginning is our Elder, and
he brings the key ; Sulphur with Salt and Mercury will
give wealth.   If you see nothing here, there is no reason
why you should search further ; so shall you be blind even
in the midst of light.'   According to Pernety,[44] Vieillard,
or Elder, is a term sometimes applied in alchemical writings to the sulphur or mercury of the philosophers, and
the ' Elder renewed in his youth ' refers to the ' first
matter ' of sulphur or gold.

The emblem [45] is closely related to the large illustration given in Mylius' *Philosophia Reformata* and reproduced in Plate 58.

§ 4. A VARIANT OF THE ' VIRIDARIUM '

Stolcius' *Viridarium* has now been examined in sufficient detail to lay bare its significance as a pictorial encyclopaedia of seventeenth-century alchemy.   It is also the
most comprehensive repository of the Frankfurt Emblems.
At the same time, it is not the only publication of its kind,
for a somewhat similar collection, attributed at the head
of the title-page to Michael Maier, was issued at Frankfurt by Herman von Sand as late as 1688, under the title
Viridarium Chymicum, or *Chymisches Lust-Gärtlein* (' Little
Pleasure Garden of Chymistry ').   This contains [46] the

first sixty-eight emblems of Stolcius' *Viridarium*, with
the omission of the illustrations from *Lusus Serius* (p. 257)
and *Septimana Philosophica* (p. 259), and of all the ' chosen
Heroes ' of the *Aureae Mensae* except Thomas Aquinas
and Michael Sendivogius. The remaining fifty-two
emblems are arranged in the order adopted by Stolcius,
and the accompanying epigrams are taken from the
German edition of Stolcius' work, published in 1624
(p. 256). The engravings are very inferior impressions
from the plates used by Stolcius. The only original
material in the publication is the preface, and even that is
modelled closely on Stolcius' introductory remarks. No
earlier edition appears to exist, and there is little doubt
that the name of Michael Maier was printed at the head
of the title-page without authority, in the hope of obtaining
a circulation for a work which is essentially an incomplete
and pirated edition of Stolcius' publication of 1624. Its
incompleteness was probably due to the loss of many
of the original plates during the intervening sixty-four
years.

§ 5. The Little Garden of Hermes

Stolcius' curious passion for providing alchemical engravings with versified Latin epigrams found a further
outlet in a second work [47] which Lucas Jennis of Frankfurt published for him in 1627. This was reprinted at
the end of the second volume of Manget's *Bibliotheca
Chemica Curiosa* (p. 116). It consists of a collection of a
hundred and sixty small drawings, each of which com-
memorates an alchemical worker, writer, or patron. Each
design is enclosed within double concentric circles measuring a little more than an inch in diameter; between the
outer and inner circles runs a Latin inscription, which
aims at expressing the main point of the subjoined epigram.[48]

The collection is entitled *Hortulus Hermeticus Flosculis
Philosophorum*, ' The Little Garden of Hermes, composed

of Flowerets of Philosophy in copper-plates, and set forth in brief verses; for the use of Students of Medical Chymistry as a repository of favoured things and for the refreshment of wearied Servants of Laboratories '. A verse addressed to the Chymistry-loving Reader strikes the same note as the title, for it runs thus: ' Child of the Fire-god, for the refreshment of thy weary mind, behold, this our little Garden will be open. In it thou shalt find flowers and a never-failing fountain, and shalt take thy pleasure of sight, scent, or sound.'

Between 1624 and 1627 Stolcius appears to have progressed in his relationships with Lucas Jennis, the Frankfurt publisher (p. 255), for he hails him here as ' the most accomplished and cultured Master Lucas Jennisius, my very dear friend '. Moreover, he puts forward a second anagram of his name, to wit, *Lucas Iennisius—In sinu lucis eas* (' mayest thou walk in the bosom of light '), which he expands into a complimentary Latin verse.

Plate 63 shows the designs which were allocated to some of the English alchemical philosophers. Roger Bacon's emblem contains a crowned hermaphroditic figure holding a sceptre, and accompanied by a dragon with four heads. The epigram takes the usual form of two Latin couplets, and runs as follows: ' Cause the fierce members of the fierce Dragon to become gentle, and let the Elements stand in their own equal grade. Then will the crowned one bring forth thence the sceptre to all, even Hermogenes; this will be the goal of thy work.' The four heads of the dragon symbolise the four elements; the crowned androgyne represents the conjunction of sophic sulphur and sophic mercury; bringing the sceptre to Hermogenes implies turning sophic mercury into gold. The encircling inscription states succinctly, ' Bring about the equalising of the elements and you will have the Magisterium '. It is interesting that the epigram which Stolcius assigned to Bacon in the *Viridarium Chymicum* (p. 259) expresses the same central idea.

THE GARDENS OF HERMES

Thomas Norton's epigram repeats a familiar alchemical notion (p. 130): ' The matter of our Stone is thought a thing of no value, so that no one would pick it up if it were found on earth. Yet the hand of God Most High has put into it, in its hiding-places, whatever this world possesses.' The inscription summarises the statement in the words: ' Our material is stuff of no price or value; whoever comes across it hardly troubles to pick it up '.

The author exercises a catholic taste in choosing subjects for his designs and epigrams; for besides most of the usual, and some unusual, ' alchymical artists ', his collection includes—presumably in an honorary capacity — such personages as Hercules, Mahomet, and Dante. Between them, the writers of the alchemical era founded and built up a remarkable ' chymic choir ' (p. 258), which included prophets, priests, patriarchs, kings, and other heroes of antiquity, both real and imaginary. Thus, to select a few typical names, Moses was included because of his convincing manipulation of the golden calf; Miriam,[49] his sister, was added on account of her supposed invention of the kerotakis (p. 15) and water-bath; Tubal-Cain was welcomed as the metallurgical expert; Cleopatra, ' Queen of Egypt ', a second and rather disturbing woman member, was elected because of her supposed invention of the alembic (p. 15) and her study of the solvent action of vinegar on pearls; Jason was admitted as the first gold-maker; and Hermes, the perpetual president of this strange alchemical society, must have been a standing embarrassment to the publication committee by reason of his reputed 36,000 original contributions to the literature! According to some of these writers, the *Song of Solomon* was an alchemical treatise, and alchemy was said to owe its name to Noah's son, Shem or Chem. No idea was too wild to find acceptance by such naïve and credulous expositors, some of whom probably looked to the Ark as the original alchemical laboratory! Their intriguing conceits had at

275

least the merit of creating a few pools of light in an illimitable ocean of misty monotony.

Besides laureating a comprehensive selection of standard alchemical figures, Stolcius commemorated certain vague characters, such as ' Medera, the Female Alchymist ', and ' Senior, the Philosopher ' (p. 268). Of the latter, he says: ' O Moon, the Spirit exalts thy child to the glittering stars, to drink in the powers of the ether. Therefore, being thus raised on high upon starry shoon, he will excel, dear Mother, thy nature.' The corresponding inscription summarises this epigram in the words: ' The generation of the Son of the Moon exceeds all his ancestry '. Stolcius thus indicates a relationship between Senior (or the Elder) and the Moon (or sophic mercury); and the epigram appears to convey the same idea as the last illustration of the *Mutus Liber* (p. 159).

A detailed examination of the hundred and sixty designs, inscriptions, and epigrams contained in this ' Little Garden of Hermes ' serves to convey an idea of the glowing enthusiasm and unending patience of alchemical workers and writers. A general view of the various works which have now been considered shows, in addition, that the alchemists of the period possessed a strictly limited number of ideas and symbolic devices, which they circulated in slightly differing forms from one publication to another. De Givry [50] remarks upon ' a powerfully expressive symbolism transmitted by the alchemists from age to age as a precious trust, which is so unalterable in its forms that a person uninitiated in alchemy may at once recognise a work, picture, or object of alchemic significance without knowing its meaning '. Undoubtedly, the fidelity of transmission is remarkable, as are also the similarities between Occidental and Oriental alchemy; but it may be argued that the repetition was largely a consequence of the paucity of alchemical ideas and the lack of essential progress which marked the alchemical era. Added to this, the alchemists recognised no laws of copyright or priority, and they borrowed freely from the fund of published material, which they seem to have regarded as the common property of those privileged citizens of the ' Republic of Chymistry ' who called themselves the ' sons of Hermes '.

§ 6. Epilogue

Francis Bacon wrote, in *De Augmentis Scientiarum*: ' Alchemy may be compared to the man who told his sons that he had left them gold buried somewhere in his vineyard; where they by digging found no gold, but by turning up the mould about the roots of the vines, procured a plentiful vintage. So the search and endeavours to make gold have brought many useful inventions and instructive experiments to light.'

This is a true saying; but, as our brief wanderings in the vasty fields of alchemy have shown, alchemy was something more than an ancient art which gave rise after long centuries to the modern science of chemistry. It was a complex and indefinite aggregation of chemistry, philosophy, religion, occultism, astrology, magic, mythology, and many other constituents. It was a strange blend of logical thinking and mystical dreaming, of sound observation and wild superstition, of natural and moral ideas, of objective facts and subjective conclusions.

The sympathetic student in search of the true significance of alchemy cannot but absorb something of the romance and wistfulness of the age-long struggle of this ' Divine Art ' after the unattainable; nor can he remain unmindful of the immemorial traditions of alchemy's daughter, chemistry. ' Pray, read, toil—and thou shalt find ', was the guiding motto of the adepts; and, at its best, alchemy was a prayerful search after truth, in the light of that principle of the unity of matter which has been rediscovered by modern science. Let us go forward in the hope that time by degrees will bring forth a corresponding unity of heart which will act as an elixir of life in the regeneration of the nations.

From *Mutus Liber*, Manget, 1702.

APPENDIX

THE MUSIC IN 'ATALANTA FUGIENS'

By F. H. Sawyer, M.A., B.Mus.

Lecturer in Music in the University of St. Andrews

THE term ' Fuga ' as applied to the musical settings of the Epigrams in *Atalanta Fugiens* is used in its earliest sense, and the work so described bears little resemblance to the highly developed fugues of a later date. The ' Fuga per Canonem ' of the sixteenth century was a work in which one voice was followed, after an interval of time, by another voice which imitated more or less strictly the preceding voice. That is to say, the second voice followed the flight of the first voice according to rule. In course of time, groups of imitating voices were separated from one another by means of episodes and a more extended work was evolved, known as the Fugue, whilst the earlier kind was termed the Canon.

The simplest type of this early Canon is to be found in the Rounds sung by children.

Though not of the primitive type, Maier's Fugues are a species of Canon, one moreover of the most rigid and mechanical kind ever devised. During the sixteenth century it was common practice for composers to base a Mass or Motett on some well-known Folk Song, as, at a later date, Bach based many of his Cantatas on Chorales. The melody was given in an augmented form to one of the voices, whilst the other voices were woven above, below, or around it. If, as in many cases, the parts were free to move without restriction, the composer could, and frequently did, write really effectively. If, on the other hand, the added parts were to be in Canon the composer had little freedom of movement. He was trebly bound, for, not only must the two parts in Canon agree with one another, but each and both together must agree with the given theme or Canto Fermo as it was styled. The procedure was of the nature of a musical jig-saw puzzle, and even at its best, as in Bach's Variations on ' Vom Himmel hoch ', it is the amazing command of technical resource shewn by Bach which impresses rather than the music itself.

In works of this kind, the main point to be emphasised is the fact that the theme used as the Canto Fermo was well known, otherwise it could have little interest for the listeners and, for all they knew, it might simply be a free part.

In Maier's Canons, the two parts styled ' Atalanta Fugiens ', and ' Hippomenes Sequens ' are invariably in Canon, whilst the part ' Pomum Morans ' is a Canto Fermo used throughout the series with little variation.   It is therefore not unreasonable to

16   FUGA II. inQuinta,infrâ.

Sein Säugmutter ist die Erden.

Atalanta Fugiens.

Romu   lus hir   ta lu   pæ presliffe sed ubera capræ

Jupiter & di   ctis fer   tur   adesse fides.

Hippomen. Sequens.

Romu lus hir   ta lu   pæ presliffe sed ubera capræ

Jupiter &   di   ctis fer   tur adesse fides.

Pomum Morans.

Romulus hirta lupæ presliffe sed ubera capræ

Jupiter & dictis fertur adesse fides.

assume that this theme may be a varied form of some tune having alchemical associations, and the fact of its constant use may be taken as symbolical of the elusive formula for which the alchemist ceaselessly sought.

As in sixteenth-century notation the musical accent followed the verbal accent, the music was generally unbarred.   Bar lines have been inserted in the transcriptions which follow for the sake of convenience, but it should be clearly understood that they possess no accentual significance and that the music should be sung without any reference to them.

Taken as a whole, the Canons possess little musical value. Maier, in setting out to write fifty Canons against the same Canto Fermo, undertook one of the most thankless tasks with which a composer could be faced, and it is not a matter for surprise that but

188 Fuga XLV. in 3. fup. vertendo Baf. & incip. ab initio in clave d.

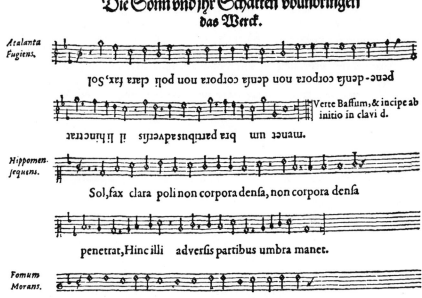

Die Sonn vnd ihr Schatten vollnbringen
das Werck.

Atalanta Fugiens.

pene-denfa corpora non denfa corpora non poli clara fax, Sol

Verte Baffum, & incipe ab initio in clavi d.

manet um bra partibus adverfis il li hinccrat

Hippomen. fequens.

Sol, fax clara poli non corpora denfa, non corpora denfa

penetrat, Hinc illi adverfis partibus umbra manet.

Fomum Morans.

Sol, fax clara poli non corpora denfa penetrat,

Hinc illi adverfis partibus umbra manet.

few possess any musical interest. Possibly a dozen, of which four are reproduced, are tolerably effective; but many of the others are marred by such incredible crudities and clashes between the parts that it is difficult to believe that they are all by the same hand. Far better examples of this type of writing are to be found in the work of composers a century previous to Maier. The work at its best can only be regarded as a musical curiosity, being one of the most extended examples of Canon against Canto Fermo in existence.

PRELUDE TO CHEMISTRY

In fairness it has to be admitted that the work appears to have been insufficiently revised, as Maier himself admits, but, as this admission only refers to the later Canons, it does not affect the errors found in the earlier ones.

Whether the Canons were ever intended to be sung whilst alchemical experiments were being carried out cannot be determined with any degree of certainty. The actual bearing of the words of the Epigrams on such experiments is by no means clear, as in no case do they suggest Invocation or Incantation. It is also difficult to believe that singers possessing the necessary musical knowledge and experience to sing them could be found amongst the laboratory assistants of Maier's time; it certainly would not be possible to find such assistants to-day. It can only be suggested that Maier prevailed on some of his friends to sing the Canons at appropriate moments during his experiments; but it is to be feared that the result can scarcely have justified the time and labour employed in preparing them.

FUGA II

THE EARTH IS THE NURSE OF THE PHILOSOPHER'S STONE

Romulus hirta lupae pressisse, sed ubera caprae
Jupiter, & factis, fertur, adesse fides . . .

We are told that Romulus sucked a she-wolf's shaggy udders,
but Jupiter a she-goat's, and that the tale is true . . .

PRELUDE TO CHEMISTRY

FUGA XXVII

HE THAT STRIVETH TO ENTER THE PHILOSOPHER'S ROSE-GARDEN WITHOUT THE KEY IS LIKENED UNTO A MAN THAT WOULD WALK WITHOUT FEET

Luxuriat Sophiae diverso flore Rosetum,
Semper at est firmis janua clausa seris . . .

Wisdom's Rose-garden flourishes with blossoms of divers kinds; but the door thereof is ever closed with fast bars . . .

FUGA XVIII

FIRE LOVES TO FIRE, BUT GOLD IT MAKETH NOT;
GOLD MAKETH GOLD, BUT FIRETH NE'ER A JOT

Si quod agens fuerit naturae, mittit in orbem
Vires atque suas multiplicare cupit . . .

Whatever active principle there is in nature, it sends out its
force in all directions and loves to multiply the same . . .

287

PRELUDE TO CHEMISTRY

FUGA XLV

THE SUN AND HIS SHADOW COMPLETE THE WORK

Sol fax clara poli, non corpora densa penetrat,
Hinc illi aversis partibus umbra manet . . .

The Sun, the bright torch of the heavens, does not pierce
opaque bodies; hence his shadow remains on the side away from
him . . .

HYDROGENESIS

Aëre cum crasso tenuis componitur aër;
Adde Iouis fulmen, fit sonus, unda manet.

With a dense gas a thin gas is mixed; pass the electric spark, and pop! you have water.

$$2H_2 + O_2 = 2H_2O.$$

GLOSSARY

This deals mainly with obsolete words and forms. Alchemical terms explained
in the text are given in the Index.

Abowte, about
Acuate, sharp, acid
Admirable, wonderful
Afore, before
Agen, again, against
Agrevyd, aggrieved
Alconomie, alchemy
Alembic, the upper part of a still
  (see *Helm*)
Alym, alum
Ameus, amiably, sweetly
Argent-vive, quicksilver, mercury;
  vivified, or mobile, silver
Argoyle, argol, crude cream of tartar
  (called *argaile* by Ben Jonson)
Athanor, furnace (Arabic, *al-
  tannur*)
Auctor, author
Auntient, ancient
Aurum, gold

Bain-marie, usually water-bath, but
  also sand (or ash) -bath
Balneo, bath
Barm, yeast
Been, are
Bemys, beams
Ben, are
Bifore, before
Bles, bliss
Blew, blue
Bloe, blue
Bloud, blood
Bodi, body
Boks, books
Brede, breadth
Brennyng, burning

Brent, burnt
Brimstoon, brimstone, sulphur
Broulleryes, misunderstandings,
  confusions

Calx vive, quicklime
Causen, cause
Cered, waxed
Chedyng, shedding
Chorle, churl
Claies, clays
Cleped, called
Cler, clear
Cley, clay
Cole, coal
Condites, conduits, water-channels
Covetise, covetousness
Curious, careful

Deceipt, deceit, deception
Dede, dead
Dele, deal, trial
Dong, dung
Dychyd, ditched
Dyd, did
Dyscevyd, deceived
Dyvers, divers, different

Eche, each
Eek, also
Egrimoigne, agrimony
Endued, endowed
Erbe, herb
Evyn, eventide
Ewel, evil
Ey, egg

Fals, false

Fatygate, fatigued

Fauchion, falchion: a short, sickle-shaped sword

Federis, feathers

Fetteth, fetches

Fimus equinis, horse-dung. According to *A Chymicall Dictionary explaining Hard Places and Words met withall in the Writings of* Paracelsus, *and other obscure Authours*, London, 1650, ' *Fimus equinis* is a digestion made any way, either by Horse-dung, or warm ashes, or water '

Flower of the Sun, the Philosopher's Stone; the Sunflower, or Marigold of Peru

Folys, fools

Fro, from

Glayre, white (also *gleire*)

Gluten, according to *A Chymicall Dictionary* (see under *Fimus equinis*), ' Gluten is any viscous matter in the body '. Pernety—ref. (II 53), 251 below—regards ' gluten of the eagle ' (*Gluten aquilae*) as synonymous with ' virgin's milk ' (*Lac virginis*)

Goon, go

Grase, graze

Grett, great

Griffon, griffin

Gryphon, griffin

Haunt, use, frequent

Heer, hear

Heggyd, hedged

Helm, the upper part of a still, from its likeness to a helmet; also *alembic*, or *limbeck*. The lower part was called the *cucurbit*, or *gourd*, also from its shape.

Hem, them

Her, here

Hertis, hearts

Hett, heat

Hir, her

Hit, it

Hors, horse

I- A.S. *ge-*, sign of the past participle passive; also *Y-*

I-blesset, blessed

I-callet, called

Insculped, engraved

I-wys, certainly, verily

Jourdayne, Jordan

Kerke, church

Lawrer, bay-tree, bay-laurel (*Laurus nobilis*), the laurel of the ancients and of poets

Leape, crack, spring

Leaper, leper

Lede, lead

Leed, lead

Lente, slow, sluggish (of a fire), gentle

Let, hinder

Levvis, leaves

Lewde, unskilful, bungling, ignorant

Limus deserti, slime of the desert

List, like

Lore, learning, knowledge

Lunayrie, lunary, moonwort, *Botrychium Lunaria*. According to *A Chymicall Dictionary* (see under *Fimus equinis*), ' Lunaria is the Sulphur of Nature '

Lust, pleasure

Lyke, like

Magnesia or *magnetia*, an indefinite term, including pyrites, magnetite, pyrolusite, and possibly magnesia. According to *A Chymicall Dictionary* (see under *Fimus equinis*), ' *Magnesia* is commonly taken for a Marcasite, but that which is artificiall is melted Tinne into which is put Quicksilver, and both mixt into a brittle matter, and white masse '

Markasits, pyrites
Martagon, same as *Lunayrie*
Materes, materials
Meddes, midst, middle
Medled, mixed, mingled
Merds, dung, excrement
Mickle, much
Mides, amidst
Mizerion, Mezereon (*Daphne Mezereum*)
Mo, more
Moche, much, many
Modre, mother
Moe, more
Mon, moon
Mowte, may, might
Myrham, myrrh

Namys, names
Notabil, notable
Nowbelson, rose noble

On, one
Onely, only
Ornate, decked, ornamented
Owre, our
Owte, out

Parfet, perfect
Payne, labour, care. *Sche dyd her payne*, She took pains
Peni, penny
Pere, peer, fellow
Plumet, plummet
Poudres, powders
Proheme, preface

Rede, red
Reede, red
Renneth, runs
Reparel, apparel
Roste, roost
Rote, root
Rownde, round

Runnett, rennet

Saine, say
Sal armoniak, sal ammoniac, ammonium chloride (also called *sal alacoph*)
Salt tartre, calcined argoyle (*q.v.*), that is, potassium carbonate
Sandifer, scum from molten glass
Sche, she
Schene, shine
Schynith, shineth
Segges, sedges
Sekeneys, sickness
Sent, scent
Silvre, silver
Slyme, slime
Snayle, snail
Sorowe, sorrow
Spagyric, iatro-chemical
Spyrytts, spirits
Stale, stole
Sterrs, stars
Stode, stood
Stoon, stone
Stremys, streams
Strooke, stroked
Suscitability, ability to become quick, or alive (Ben Jonson)
Suttill, subtle
Sweght, sweet

Tarie, tarry, wait
Tast, taste
Than, then
The, thee
Ther on, thereon
Therthe, the earth
Tought, taught
Travel, work
Trowe, true, trust
Tuching, touching
Tus, frankincense
Tutits, oxide or carbonate of zinc (also *tutia*)
Tweye, two
Tyned, kindled

Ungued, ungual
Unparfyt, imperfect
Unpittiful, pitiless
Unslekked lyme, quicklime

Vaile, veil
Velame, vellum
Vervaine, vervain, vervein (*Verbena officinalis*): a plant used in love-philtres, charms, etc.
Vitriall, vitriol: a glistening, crystalline body

Warke, work
Weer, were
Well, rise, spring
Wellid, welled

Wenyng, thinking
Werk, work
Wexsyth, waxeth, grows
Whilome, once, formerly
Wote, know, knew
Wyche, which
Wyfe, wife
Wys, see *I-wys*

Y- A.S. *ge-*, sign of the past participle passive (also *I-*)
Yerne, iron
Y-lyche, alike
Ynperfection, imperfection
Yron, iron
Ys, is
Yt it

BIBLIOGRAPHY AND NOTES

THE bibliography is restricted to literature quoted in the text of this book. The first number given refers to the same number printed in the text of the Chapter concerned. Volume numbers are printed in heavy type. As a rule, in order to avoid repeating titles, the references are carried back to the original entry: for example, in reference 34 below, ' Hopkins, (I 22), 120 ' means page 120 of the work by A. J. Hopkins, of which details are given in entry 22 of Chapter I.

THE PROHEME

1 G. Sarton, *Science*, 1917, **45**, 284.

CHAPTER I

1 See lines 2 to 5 of John Gower's poem *Concerning The Philosopher's Stone*, which is printed on pp. 368-373 of E. Ashmole's *Theatrum Chemicum Britannicum*, London, 1652. The title-page of this important collection of English alchemical poems runs as follows:
   Theatrum Chemicum Britannicum. *Containing* Severall Poeticall Pieces of our Famous *English Philosophers*, who have written the *Hermetique Mysteries* in their owne Ancient Language. Faithfully Collected into one Volume, with Annotations thereon, *By* Elias Ashmole, *Esq. Qui est Mercuriophilus Anglicus.* The First Part. *London*, Printed by *J. Grismond* for Nath: Brooke, at the Angel in *Cornhill. MDCLII.*
2 W. Salmon, *Polygraphice*, 8th edn., London, 1701, 477. The ' semi-metals ' are here given as antimony, cinnabar, tin-glass, and zink.
3 J. von Liebig, *Familiar Letters on Chemistry*, London, 1859, 54.
4 M. A. Atwood (Mrs), *A Suggestive Inquiry into the Hermetic Mystery*, 3rd edn., Belfast, 1920, p. 26 of Introduction.
5 G. de Givry (trans. J. C. Locke), *Witchcraft, Magic and Alchemy*, London, 1931, 374, 350. For a somewhat similar view, see A. E. Waite, *Lives of Alchemystical Philosophers*, London, 1888, 33 *et seq.*

6   Lenglet Dufresnoy, *Histoire de la Philosophie Hermétique*, 3 vols., Paris, 1742, **1**, 1.

7   E. J. Holmyard, *The Great Chemists*, London, 1928, 2.

8   See the contribution by A. Lucas to *The Analyst*, 1933, **58**, 654, on 'Ancient Egyptian Materials and Industries about 1350 B.C.'

9   J. R. Partington, *Origins and Development of Applied Chemistry*, London, 1935, 5, states: 'The hieroglyphic name for Egypt, supposed to represent a heap of charcoal, a crocodile's tail or a piece of fish skin, is qemi or chemi, with the meaning black—the black land; chemia is a rare form'.

10 Y. Y. Ts'ao, *Science*, 1933, **17**, 31-54 (Science Society of China, Shanghai); *Journal of Chemical Education*, 1934, **11**, 655.

11 A. Waley, *The Travels of an Alchemist: the Journey of the Taoist Ch'ang-Ch'un from China to the Hindukush at the Summons of Chingiz Khan* (translation with introduction), London, 1931.

12 A. Waley, (I 11), 12; G. Sarton, *Introduction to the History of Science*, Baltimore, 1927, **1**, 355; T. L. Davis, *Journal of Chemical Education*, 1934, **11**, 517; Lu-Ch'iang Wu and T. L. Davis, *Proceedings of the American Academy of Arts and Sciences*, 1935, **70**, 221.

13 A. Waley, 'Notes on Chinese Alchemy', in the *Bulletin of the School of Oriental Studies, London Institution*, 1930, **6**, part 1; see also O. S. Johnson, *A Study of Chinese Alchemy*, Shanghai, 1928; and T. L. Davis, *Isis*, 1932, **18**, 224.

14 R. N. Bhagvat, *Journal of Chemical Education*, 1933, **10**, 661.

15 T. L. Davis, in an introduction to 'An Ancient Chinese Alchemical Classic: Ko Hung on the Gold Medicine and on the Yellow and the White, the fourth and sixteenth chapters of Pao-p'u-tzŭ, translated from the Chinese by Lu-Ch'iang Wu', in *Proceedings of the American Academy of Arts and Sciences*, 1935, **70**, 221. See also T. L. Davis, *Isis*, 1932, **18**, 213.

16 *An ancient Chinese treatise on alchemy entitled Ts'an T'ung Ch'i*, written by Wei Po-yang about 142 A.D., now translated into English by Lu-Ch'iang Wu, with an introduction and notes by Tenney L. Davis, in *Isis*, 1932, **18**, 210-289.

17 Partington, (I 9).

18 E. J. Holmyard, *Nature*, 1929, **123**, 520.

19 G. F. Rodwell, *The Birth of Chemistry*, London, 1874, 27.

20 Bhagvat, (I 14), 659.

21 M. M. Pattison Muir, *The Story of Alchemy and the Beginnings of Chemistry*, London, 1902, 52. See also R. Steele on 'Alchemy', in *Shakespeare's England*, Oxford, 1916, **1**, 463.

22 A. J. Hopkins, *Alchemy Child of Greek Philosophy*, New York, 1934, 28.

23 *Idem*, 32.

24 T. L. Davis, *Annals of Medical History*, 1924, **6**, 280.

25 Hopkins, (I 22), 69.

26 *Idem*, 42.

27   *Idem*, 49.
28   *Idem*, 92.
29   *Idem*, 115.
30   *Idem*, 117.
31   *Idem*, 123.
32   *Idem*, 176.
33   For a useful bibliography dealing with Geber, see E. J. Holmyard (ed.), *The Works of Geber, Englished by Richard Russell*, 1678, London, 1928, xxii; see also Hopkins, (I 22), 134 *et seq.*
34   Hopkins, (I 22), 120.
35   Davis, *Journal of Chemical Education*, 1935, **12**, 5; see also Davis, *Isis*, (I 15), 216.
36   Bhagvat, (I 14), 666.
37   Davis, (I 15), 245.
38   Waley, (I 11), 22.
39   Davis, *Isis*, (I 15), 222.
40   G. Chaucer, *The Canones Yeomans Tale*, in *The Canterbury Tales*, ed. A. Burrell, Everyman's Library, London, 421.
41   Atwood, (I 4), 32.
42   Hopkins, (I 22), 185.
43   The quotation is from the English translation, *The Mirror of Alchimy*, of 1597; see also T. L. Davis, *Journal of Chemical Education*, 1931, **8**, 1945.
44   Reproduced in D. Stolcius' *Viridarium Chymicum*, Frankfurt, 1624, from M. Maier's *Symbola Aureae Mensae Duodecim Nationum*, Frankfurt, 1617.
45   B. Jonson, *The Alchemist*, London, 1612: 'The King's Library' edn., London, 1903, 45: Act 2, sc. 1, l. 365.
46   A. E. Waite (ed.), *The Hermetic and Alchemical Writings of Aureolus Philippus Theophrastus Bombast, of Hohenheim, called Paracelsus the Great. Now for the first time faithfully translated into English*, 2 vols., London, 1894, **1**, 125. This is a translation of the Latin edn., Geneva, 1658.
47   Mercury and sulphur are often incorrectly associated in modern literature with the soul and spirit, respectively: see, for example, T. Thomson, *The History of Chemistry*, London, 1830, **1**, 157; J. E. Mercer, *Alchemy*, London, 1921, 218.
48   Waite, (I 46), **1**, 149.
49   *The Last Will and Testament of Basil Valentine*, London, 1671, 311.
50   A. Poisson, *Théories & Symboles des Alchimistes*, Paris, 1891, 17.
51   P. Lacroix, *Science and Literature in the Middle Ages*, London, 1878, 190.
52   Poisson, (I 50), v.
53   T. Thomson, *The History of Chemistry*, 2 vols., London, 1830, **1**, 146.
54   Lacroix, (I 51), 192.
55   H. C. and (Mrs) L. H. Hoover, *De Re Metallica*, London, 1912, xxvii, footnote 12. The full title of this work is as follows:

Georgius Agricola, *De Re Metallica*, translated from the first Latin edition of 1556 with Biographical Introduction, Annotations and Appendices upon the Development of Mining Methods, Metallurgical Processes, Geology, Mineralogy & Mining Law from the earliest times to the 16th Century, by H. C. Hoover and L. H. Hoover, London, 1912.

56　R. Boyle, *The Sceptical Chymist: or Chymico-physical Doubts and Paradoxes*, London, 1661. A modern reprint of this celebrated work is available in Everyman's Library.

57　*Idem*, 343, 350.

58　Hopkins, (I 22), 225.

59　See, for example, J. Read, *A Text-Book of Organic Chemistry*, 2nd edn., London, 1935, 20 *et seq.*

60　Davis, (I 15), 221.

61　J. Priestley, *Experiments and Observations on Different Kinds of Air*, 3 vols., London, 1774–1777.

62　G. Wilson, *Life of Cavendish*, London, 1851, 186.

CHAPTER II

1　*Bloomefields Blossoms*, printed by Ashmole, (I 1), 308.

2　C. P. Bryan, *The Papyrus Ebers*, London, 1930, 9; see also A. C. Wootton, *Chronicles of Pharmacy*, 2 vols., London, 1910, 1, 36.

3　R. C. Thompson, *On the Chemistry of the Ancient Assyrians*, London, 1925.

4　See (I 15).

5　For details of early Greek MSS. on alchemy, see *e.g.*, J. Bidez *et al.*, (II 12); J. C. Brown, *A History of Chemistry from the Earliest Times*, 2nd edn., London, 1920, 45; J. M. Stillman, *The Story of Early Chemistry*, New York, 1924, 78; Sarton, (I 12); Hopkins, (I 22); and the works of M. Berthelot mentioned in the text.

6　Jonson, (I 45), 32: Act 2, sc. 1, l. 80.

7　M. Berthelot, *Introduction à l'Étude de la Chimie des Anciens et du Moyen Age*, Paris, 1889, 152; Brown, (II 5), 45.

8　E. J. Holmyard, *Makers of Chemistry*, Oxford, 1931, 35, 39.

9　L. Figuier, *L'Alchimie et les Alchimistes: essai historique et critique sur la philosophie Hermétique*, 3rd edn., Paris, 1860, 7.

10　J. H. Bridges (ed.), *Opus Majus*, Oxford, 1897, 2, 167.

11　Hopkins, (I 22), 64, 192.

12　J. Bidez *et al.*, *Catalogue des manuscrits alchimiques grecs;* (Mrs) D. W. Singer, *Catalogue of Latin and Vernacular Alchemical Manuscripts in Great Britain and Ireland dating from before the XVI Century*, 3 vols., Brussels, 1928–1931.

13　British Museum, Sloane MS. 1754, ff. 48-50; Singer, (II 12), 178.

14　British Museum, Sloane MS. 353, ff. 56ᵛ-61ᵛ; Singer, (II 12), 153.

15　Some of the drawings in this MS. appear, with slight differences in detail, in *La Clef de la Grand Science sur l'ouvrage philosophique inconnu jusqu'à présent* (Bibliothèque de l'Arsenal, MS. 6577), and in Barchusen's *Elementa Chemiae* (Leyden, 1718).   The drawings dealing with the story of the Creation include coloured versions of the 8 illustrations given in Figura III of *Janitor Pansophus* (' The All-Wise Doorkeeper '), at the end of the *Musaeum Hermeticum* (Frankfurt, 1678).   One of the quaintest of the coloured illustrations of the Creation in this MS. depicts the birth of Eve, with the serpent as a spectator.

16   G. Sarton, *Introduction to the History of Science*, Baltimore, 1931, **2**, 1043.
  Among early printed versions of the *Summa Perfectionis* are two editions issued by Johannes Grüninger at Strassburg in 1529 and 1531.   The St. Andrews collection contains a copy of the latter, consisting of 60 leaves, numbered on one side only and entitled *Geberi Philosophi ac Alchemistae Maximi, de Alchimia libri Tres* (' Three Books of Alchemy, by Geber, the great Philosopher and Alchemist ').   This contains 4 large woodcuts, and 6 smaller ones, all of which are distinct from those of the Berne edition of 1545. Some of the smaller woodcuts were taken from Brunschwick's *Buch zu Distillieren* (see p. 75), also printed by Grüninger at Strassburg and issued in editions bearing 1519 and other dates. The woodcuts from these editions of Geber were also used by Grüninger in Ulstad's *Coelum Philosophorum* (' Philosophers' Heaven '), published at Strassburg in 1528, etc.   Illustrations were passed from one work to another with great liberality by alchemical writers and publishers.

17   J. Ruska, *Annales Guébhard-Séverine*, Neuchâtel, 1934, **10**, 410.   For the complex bibliography of the Geber texts, see *inter al.*, Dufresnoy, (I 6), **3**, 33;   J. Ferguson, *Bibliotheca Chemica*, Glasgow, 1906, **1**, 18, 302;   Holmyard, (I 33); Hopkins, (I 22).

18   J. Ferguson, *Bibliotheca Chemica*, 2 vols., Glasgow, 1906, **1**, 393.

19   J. Ruska, *Tabula Smaragdini*, Heidelberg, 1926, 193.

20   E. J. Holmyard, *Nature*, 1923, **112**, 526.

21   Ruska, (II 19), 121.

22   Rodwell, (I 19), 62.   See also Thomson, (I 51), **1**, 10; *Musaeum Hermeticum*, Frankfurt, 1678, final plate (No. 4); A. E. Waite, *The Hermetic Museum, Restored and Enlarged*, London, 1893, **2**, 243.

23   J. Lacinius (ed.), *Pretiosa Margarita Novella de Thesauro, Ac Pretiosissimo Philosophorum Lapide*, Venice, 1546.   See also (II 24) and (II 27).

24   The existence of this edition of *Pretiosa Margarita Novella* has been often overlooked;   it is unmentioned by Ferguson or Lenglet Dufresnoy.   It was published by Ziletti, who introduced a new titlepage with his device and reprinted the first 8 leaves.   Apart from this, the edition is exactly like the original Aldine issue of 1546.

25   J. Ruska, *Annales Guébhard-Séverine*, Neuchâtel, 1934, **10**, 417.

26 See Ferguson, (II 18), **2**, 478; A. E. Waite (ed.), *The Turba Philo-sophorum: or Assembly of the Sages*, London, 1914.

27 From the English translation of *Pretiosa Margarita Novella*, issued by A. E. Waite, London, 1894, under the title: *The New Pearl of Great Price. A Treatise concerning the Treasure and Most Precious Stone of the Philosophers* (p. 93).

28 A. Poisson, *Nicolas Flamel*, Paris, 1893, 83.

29 Lacroix, (I 51), 196.

30 See A. E. Waite, *Lives of Alchemystical Philosophers*, London, 1888, 98.

31 Plate 7 is reproduced from A. Poisson's *Nicolas Flamel* (Paris, 1893). An almost identical print occurs in G. de Givry's *Witchcraft, Magic and Alchemy* (London, 1931). Eirenaeus Orandus' *Nicholas Flammel* (London, 1624) contains a crude engraving omitting all the figures above the arch; this engraving is repeated in sections later in the book. A later reproduction of Flammel's ‘ hieroglyphicks ’ may be found in W. Salmon's *Medicina Practica*, London, 1707.

32 Bibliothèque Nationale, Paris, fonds français 14765; this drawing has been reproduced in colour by de Givry (I 5).

33 Eirenaeus Orandus, *Nicholas Flammel*, London, 1624, 95 *et seq.*

34 *Idem*, 21.

35 *The Golden Calf, Which the World Adores, and Desires*, faithfully Englished from the Latin of J. F. Helvetius, London, 1670, 45 *et seq.*

36 See Figuier, (II 9), 217; Lacroix, (I 51), 196; Waite, (II 30), 110; Poisson, (II 28), 90; Ferguson, (II 18), **1**, 280, **2**, 491.

37 The translations are taken, with several slight amendments, from an anonymous modern English version of *Splendor Solis*, with notes, etc., by J. K., London, n.d. [1921].

   It may be noted that *Splendor Solis* was also printed as the first of a series of tracts forming a volume of 701 pages (in the St. Andrews collection) under the title *Aurei Velleris oder Der Güldin Schatz und Kunstkammer. Tractatus III.* The title-page bears the date 1600, but no indication of place of issue or printer is given. The contents differ markedly from those of the *Aureum Vellus* of Rorschach, etc., and particularly in the inclusion of Basil Valentine's *Von dem grossen Stein der Uhralten* and *Zwölff Schlüssel*. These additions suggest that the publication is a spurious one, deliberately antedated (p. 184); indeed, it may be one of the ‘ pirated ’ editions of which Johann Thölde complained (p. 194). This version of *Splendor Solis* contains all the mistakes of the Rorschach edition, from which it appears to have been copied: the standard text of this tract is that of the Harley MS. 3469.

38 Lacroix, (I 51), 185. See also Steele, (I 21), 462.

39 O. Guttmann, *Monumenta Pulveris Pyrii*, London, 1906.

40 Hoover, (I 55), 546.

41 See (I 55).
42 Hoover, (I 55), 40.
43 *Idem*, 217.
44 Figuier, (II 9), 255.
45 H. Khunrath, *Amphitheatrum Sapientae Æternae*, Hanover, 1609, *Gradus Tertii*, 67.
46 Waite, (II 30), 159.
47 A reproduction of Khunrath's Alchemical Citadel is given by de Givry, (I 5), 348.
48 A. Tennyson, *Morte D'Arthur*.
49 de Givry, (I 5), 350.
50 Fig. 8 is reproduced from J. J. Manget, *Bibliotheca Chemica Curiosa*, 2 vols., Geneva, 1702, **2**, 287; the bottom horizontal line measures about 8·6 cm. in the original engraving.   See VII 48.
51 Trismosin, (II 37), 81.
52 Ashmole, (I 1), 278.
53 A. J. Pernety, *Dictionnaire Mytho-Hermétique*, Paris, 1787, vi, xxi.
54 See T. L. Davis on ' Primitive Thinking ', in *The Technology Review*, 1929, **31**; and in the *Journal of Chemical Education*, 1935, **12**, 9.
55 Chaucer, (I 40), 425.
56 Ashmole, (I 1), 368.
57 Basil Valentine, (I 49), plate preceding p. 1.
58 Ashmole, (I 1), Prolegomena, p. vi (unnumbered).
59 A. E. Waite (ed.), *The Hermetic Museum, Restored and Enlarged*, 2 vols., London, 1893, **2**, 243.
60 Liebig, (I 3), 47.
61 Jonson, (I 45), 34, 47, 33: Act 2, sc. 1, l. 94.
62 Ashmole, (I 1), 425.
63 Basil Valentine, (I 49), 137.
64 J. E. Mercer, *Alchemy, its Science and Romance*, London, 1921, 106.
65 Ashmole, (I 1), 114.
66 *Idem*, 19.
67 Trismosin, (II 37), 15.
68 Ashmole, (I 1), 121.
69 *Idem*, 463.
70 *Idem*, 348.
71 J. Gerard, *Herball*, 2nd edn., London, 1636, chap. 89: ' The small Lunary springeth forth of the ground with one leafe like Adderstongue, jagged or cut on both sides into five or six deepe cuts or notches. . . . The root is slender, and compact of many small threddy strings. . . .
     ' *Lunaria* or small Moone-wort groweth upon dry and barren mountains and heaths.   I have found it growing in these places following; that is to say, about Bathe in Somersetshire in many places, especially at a place called Carey, two miles from Bruton, in the next close unto the Churchyard; on Cockes Heath betweene Lowse and Linton, three miles from Maidstone in Kent: it groweth

also in the ruines of an old bricke-kilne by Colchester, in the ground of Mr *George Sayer*, called Miles end: it groweth likewise upon the side of Blacke-heath, neere unto the stile that leadeth unto Eltham house, about an hundred paces from the stile: also in Lancashire neere unto a Wood called Fairest, by Latham: moreover, in Nottinghamshire by the Westwood at Gringley, and at Weston in the Ley field by the West side of the towne; and in the Bishops field at Yorke, neere unto Wakefield, in the close where Sir *George Savill* his house standeth, called the Heath Hall, by the relation of a learned Doctor in Physicke called Master *John Mershe* of Cambridge, and many other places.

' *Lunaria* or small Moone wort is to be seene in the moneth of May.' (See also M. Woodward, *Gerard's Herball*, London, 1927, 102.)

72   Ashmole, (I 1), 315.
73   Trismosin, (II 37), 29.
74   Davis, (I 15), 259.
75   Ashmole, (I 1), 99.
76   *Idem*, 120.
77   *Idem*, 90.
78   *Idem*, 187.
79   Flamel, (II 33), 82-86.
80   Jonson, (I 45), 57: Act 2, sc. 1, l. 610.
81   *Idem*, 45: Act 2, sc. 1, l. 374.
82   *Idem*, 44: Act 2, sc. 1, l. 342.
83   This design is reproduced from D. Stolcius, *Viridarium Chymicum*, Frankfurt, 1624, Figura XCV; it occurs also in *Azoth* (see p. 270 of this book).
84   H. S. Redgrove, *Bygone Beliefs*, London, 1920, 68.
85   The Book of *Genesis*, iii, 1.
86   See Brown, (II 5), 20; A. J. Hopkins, (I 22), 107.
87   Hopkins, (I 22), 32.
88   Davis, (I 24).
89   T. L. Davis, *Journal of Chemical Education*, 1935, **12**, 7.
90   Waite, (II 59), **2**, 319.
91   Ashmole, (I 1), 350
92   Basil Valentine, (I 49), **347.**
93   Ashmole, (I 1), 323.
94   Basil Valentine, (I 49), 123.
95   Singer, (II 12), **1**, frontispiece; also British Museum, Egerton MS. 845, f. 19^v.
96   Waite, (II 59), **1**, 180.
97   *Idem*, **1**, 236.
98   Ashmole, (I 1), 466.
99   *Idem*, 154.
100 de Givry, (I 5), 363.
101 Lacroix, (I 51), 190.

102   Flamel, (II 33), 34.
103   A. E. Waite, *The Brotherhood of the Rosy Cross*, London, 1924, 312.
104   Redgrove, (II 84), 141.
105   Waite, (II 59), **1**, ix.
106   This collection is preserved in the Royal Technical College, Glasgow.
      The still more comprehensive Ferguson Collection, made by John
      Ferguson, Professor of Chemistry in the University from 1874 to
      1915, is contained in the Library of the University of Glasgow.

CHAPTER III

 1   Liebig, (I 3), 53.
 2   Quoted by Atwood, (I 4), 72.
 3   Pernety, (II 53), 17.
 4   Liebig, (I 3), 78.
 5   Davis, (I 15), 224, 244.
 6   *Idem*, (I 15), 213.
 7   *Idem*, (I 15), plate 3, opposite p. 212.
 8   Quoted in *The New Pearl of Great Price*, (II 27), 348.
 9   Ashmole, (I 1), 116.
10   Ferguson, (II 18), **2**, 470.
11   Figuier, (II 9), 227.
12   Ashmole, (I 1), 449.
13   Waite, (II 59), **1**, 236.
14   Jonson, (I 45), 31; Act 2, sc. 1, l. 47.
15   Trismosin, (II 37), 79.
16   Ashmole, (I 1), Prolegomena, p. vi (unnumbered).
17   Compare Eirenaeus Philalethes, *The Secret of the Immortal Liquor
      called Alkahest or Ignis-Aqua*, included in *Collectanea Chemica*,
      London, 1893, 9.
18   F. Hoefer, *Histoire de la chimie*, 2 vols., Paris, 1866–1869, **2**, 17.
19   *The names of the Philosopher's Stone, Collected by* William Gratacolle.
      Translated into English By the Paines and Care of H. P. London,
      Printed by *Thomas Harper*, and are to be sold by *John Collins*, in
      Little Brittain, near the Church door, 1652.
20   Pernety, (II 53), 272.
21   Ashmole, (I 1), 312.
22   Mercer, (II 64), 133.
23   Waite, (II 59), **2**, 249.
24   Jonson, (I 45), 57: Act 2, sc. 1, l. 619.
25   Waite, (II 59), **1**, 180.
26   Ashmole, (I 1), 342.
27   *Idem*, (I 1), 88.
28   See H. Kopp, *Die Alchemie in älterer und neuer Zeit*, 2 vols., Heidel-
      berg, 1886, **2**, 246.
29   An interesting summary of some of the practical processes concerned

here, with references to the original alchemical literature, is given
by Poisson (I 50).

30   Dufresnoy, (I 6), **2**, 342.
31   Liebig, (I 3), 46.
32   Trismosin, (II 37), 40.
33   Chaucer, (I 40), 424.
34   Ashmole, (I 1), 189.
35   Pernety, (II 53), 99.
36   Ashmole, (I 1), 129.
37   Jonson, (I 45), 56: Act 2, sc. 1, l. 599.
38   See de Givry, (I 5), 368.
39   Waite, (II 59), **2**, 159.
40   *A Treatise written by* Alphonso *King of Portugall, concerning the
     Philosophers Stone.* Translated into English By the Paines and
     Care of H. P. London, 1652.
41   J. R. Glauber, *Works* (trans. C. Packe), London, 1689, part iii, 20.
42   Quoted in *The New Pearl of Great Price*, (II 27), 414.
43   Quoted in *The New Pearl of Great Price*, (II 27), 343.
44   Pernety, (II 53), 153.
45   Salmon, (I 2), 478, 503-509. For example (p. 505): ' The base
     Metals yield an increase according to the quantity of the *Mercury*
     which they contain, and that is the substance which is transmuted;
     all the other substances are Heterogeneous and prove Dross, which
     by the fermentative power of our *Solar* and *Lunar Sulphur*, are
     separated and cast off, (almost resembling that of Yest in Wort:)
     . . . the *immature Mercury* in the baser Metals feeds our *Solar* and
     Lunar Seeds, and makes them to grow and encrease '.
46   Davis, (I 15), 237.
47   Ashmole, (I 2), 181, 185.
48   Jonson, (I 45), 43: Act 2, sc. 1, l. 316.
49   Quoted in *The New Pearl of Great Price*, (II 27), 346.
50   Jonson, (I 45), 33: Act 2, sc. 1, l. 107.
51   Salmon, (I 2), 509.
52   Waite, (II 59), **2**, 247; see also Liebig, (I 3), 46.
53   Ashmole, (I 1), 103.
54   Jonson, (I 45), 72: Act 3, sc. 2, l. 185.
55   *Proceedings of the American Academy of Arts and Sciences*, 1935, **70**,
     251. Cf. (I 15).
56   See (I 16).
57   Ashmole, (I 1), 321.
58   Waite, (I 46), **1**, 27.
59   Ashmole, (I 1), 426.
60   Waite, (II 59), **2**, 186.
61   Davis, (I 15), 240.
62   Chaucer, (I 40), 427.
63   Ashmole, (I 1), 188.
64   Waite, (I 46), **1**, 126.

65 Ashmole, (I 1), 115.
66 Waite, (II 59), **2**, 182.
67 Davis, (I 15), 239.
68 British Museum, Harley MS. 3469.
69 Waite, (II 30), 36.
70 See, for example, de Givry, (I 5), 372.
71 Jonson, (I 45), 72: Act 3, sc. 2, l. 175.
72 Lewis Carroll, *Through the Looking-Glass.*
73 Waite, (II 59), **2**, 186.
74 Ashmole, (I 1), 421.
75 Brown, (II 5), 46.
76 Ferguson, (II 18), **1**, 29; see also Kopp, (III 28), **2**, 318.
77 Arcere, *Histoire de la ville de la Rochelle*, 1757, **2**, 384.
78 The Book of *Genesis*, xxviii, 12.
79 Salmon, (I 2), 635.
80 Jonson, (I 45), 34: Act 2, sc. 1, l. 132.
81 Davis, (I 15), 261.
82 Jonson, (I 45), 47: Act 2, sc. 1, l. 417.
83 See, for example, p. 232 of *Artephius his Secret Booke, Of the blessed
    Stone, called the Philosophers.* London.   Printed by *T. S.* for
    Tho. Walkley, and are to be sold at his Shop at the Eagle and Childe
    in Britans Bursse.   1624.   (The pagination of this book begins at
    141, being continued from that of *The Hieroglyphicall figures of
    Nicholas Flammel,* with which it is bound—see II 31.)
84 *Idem,* 207.
85 J. D. Mylius, *Philosophia Reformata,* Frankfurt, 1622, book I, part
    vii, 297.
86 Ashmole, (I 1), 342.
87 Compare Dufresnoy, (I 6), **2**, 343.
88 Ashmole, (I 1), 420.
89 Chaucer, (I 40), 422, 441.
90 Liebig, (I 3), 56.
91   *Se 'l grand' arcano ritrovar intendi,
                Ah cessa.   Il tempo e la fatica getti.
                A miglior uso si riserbi l' Arte
                Che all' altre suol esser Maestra, e duce.*
        This engraving, contained in the St. Andrews collection, is
    marked *Teniers pinx.   Fran^co Pedro sculp. apud N. Cavalli Venetiis.*
    It measures about 39 by 30 cm.   The chief figure is looking to the
    left.   A similar engraving, in the same collection (see Plate 29),
    is marked *D. Teniers pinx.   F. Basan sculp.   A Paris chez
    Basan, rüe du Foin.*   It measures about 40 by 30 cm., and is
    entitled ' Le Grimoire d'Hypocrate '.   The chief figure is looking
    to the right, and the details are similarly reversed.

CHAPTER IV

1 Waite, (II 59).

2 Jean de Meun, Meung, or Mehung, completed in the second half of
   the thirteenth century, the celebrated *Roman de la Rose*, one of the
   greatest poems of mediaeval literature. According to Sarton,
   (II 16), **2**, 932-933, the alchemical writings ascribed to John of
   Mehung are apocryphal.

3 Curiously, a closely similar title-page adorns the English version of
   Otto Tachenius' *Hippocrates Chymicus*, published at London in
   1677 by ' Nath: Crouch, at the George at the lower end of Cornhill over against ye Stocks Market '. The designs are identical,
   except that in the English work the top medallion shows four miners
   at work, and the bottom one depicts the interior of an apothecary's
   shop; these two medallions were derived from the title-page of the
   third tract in the *Musaeum Hermeticum*.

4 Waite, (II 59), **2**, 321.

5 The Latin version of the *Tripus Aureus* was reprinted in the *Musaeum
   Hermeticum* of 1678 and 1749.

6 Waite, (II 59), **2**, 70.

7 Ashmole, (I 1), 467.

8 *The Diary of Abraham de la Pryme*, Surtees Society, 1870, **54**, 104.

9 Ashmole, (I 1), 443.

10 *Idem*, 210, 465.

11 M. Nierenstein and P. F. Chapman, in an exhaustive study of the
   authorship of the *Ordinall* (*Isis*, 1932, **18**, 290-321), came to the
   conclusion that ' Maier and Ashmole were not justified in definitely
   ascribing the anonymously written *Ordinall* either to Thomas
   Norton . . . or to Thomas Norton of Bristoll '. They point out
   that no known Thomas Norton appears to fulfil the requirements.
   However, Norton is a common surname and place-name in the
   Bristol region, and there seems to be no cogent reason for supposing
   that the *Ordinall* was not written by an unidentified Thomas
   Norton of Bristol: if this were not so, the cipher would be either
   accidental or deliberately misleading, and there is no evidence
   indicating that it is either. This interesting paper provides a good
   example of the complications and difficulties which are inherent in
   the detailed study of mediaeval alchemical writings.

12 The versified Latin translation appears also in J. J. Manget's *Bibliotheca Chemica Curiosa*, Geneva, 1702, **2**, 285-309. The title-
   page of Maier's version of the *Ordinall* is embellished with an
   engraving which ostensibly portrays Thomas Norton. An impression from the same plate appeared in Maier's *Symbola Aureae
   Mensae Duodecim Nationum* of 1617, and also in Stolcius' *Viridarium Chymicum* of 1624; but in each of these works it purports
   to represent Raymond Lully. Maier played a similar trick with
   Cremer and Thomas Aquinas (see p. 170 of this book). More-

over, the engraving depicting Roger Bacon in his *Symbola* of 1617 (Plate 41 (ii) in this book) does duty for Basil Valentine in his *Tripus Aureus* of 1618 (p. 7), as an ancient commentator has testily remarked in a faded marginal note, written in Latin in the St. Andrews copy of the *Tripus*.

13 Ashmole, (I 1), Prolegomena, p. ii (unnumbered); see also Waite, (II 103), 310-339.

14 An entry in one of his note-books (1662–1669) shows that Sir Isaac Newton bought a copy of Ashmole's *Theatrum Chemicum Britannicum* in 1669 for £1 : 8 : 0.   Manuscript transcripts of parts of this book, in Newton's handwriting, are still extant.

It is stated that alchemy and theology were Newton's two abiding interests, and that surviving manuscripts on alchemy in his hand comprise some 650,000 words. 'They show him to have assimilated the whole corpus of Alchemical Literature and to have been the most learned adept of all time.   He was also a very skilful experimenter and a great part of his thirty-five years at Cambridge was spent among retorts and furnaces in the laboratory which he had built for himself.'   Humphrey Newton, his assistant from 1685 to 1690, wrote that ' especially at Spring and Fall of y[e] Leaf, . . . he us'd to imploy about 6 weeks in his Elaboratory, the Fire scarcely going out either night or day, he siting up one Night as I did another, till he had finished his Chymical Experiments, in y[e] Performance of w[ch] he was y[e] most accurate, strict, exact: What his aim might be I was not able to penetrate into, but his Pains, his Diligence, at these sett Times made me think he aimed at something beyond y[e] Reach of humane Art and Industry.'

The above information has been extracted from the *Catalogue of The Newton Papers Sold by Order of The Viscount Lymington to whom they have descended from Catherine Conduitt, Viscountess Lymington, Great-niece of Sir Isaac Newton* (Sotheby & Co., London, 1936).   The writer of the foreword to this publication, in commenting upon the last quotation, makes the following interesting observations:

' And so he did, for he had set himself to discover the Elixir of Life, and how to transmute base metals into gold.   He was more than usually secretive about these romantic pursuits.   Although so much of his life was spent in the company of Diana's Doves, chasing the Red and Green Lyons through the Twelve Gates, or elevating Mercury with the full complement of Ten Eagles, he published only one chemical paper (*De natura acidorum*) and this gave no inkling of the ultimate and magnificent object of his researches. After his appointment to the Mint, of course, any open association of his name with Alchemy would have been most indiscreet.   The rumour that the Master of the Mint could transmute copper farthings into bright golden guineas would have spread panic through the nations. . . . The Alchemy that Newton practised had more than its vocabulary in common with Mysticism, and no doubt it was by way of Alchemy that Newton entered upon the Interpretation of the Prophecies which forms so large a part of his Theological writings. These amount to more than one-and-a-quarter million words and are mostly unpublished.'

15 Among several MS. copies of Norton's *Ordinall* contained in the British Museum (15th to 17th cent.) is a fifteenth-century copy (Add. MS. 10302) which may just possibly have been Norton's own. This contains illustrations in colour, of which one (fol. 37^b) is reproduced in Plate 34; but probably Ashmole's illustrations, and certainly his characteristic borders, were derived from some other source.

An English prose translation of the Latin version of Norton's *Ordinall* was published in 1893; see Waite, (II 59), **2**, 1-57. A facsimile reproduction of Ashmole's edition has also been published: E. J. Holmyard (ed.), *The Ordinall of Alchimy*, London, 1928.

16 Ashmole, (I 1), 437. Two discrepancies occur in Ashmole's note on this cipher: (1) the word ' Tomais ' is spelt ' Tomas ', which is only allowable if ' maistryefull ' is spelt ' mastryefull '; (2) the word ' call ' is omitted from the second line.

17 M. Maier, *Symbola Aureae Mensae Duodecim Nationum*, Frankfurt, 1617, 467.

18 Ashmole, (I 1), 442.

19 *Idem*, 33.

20 Waite, (II 59), **2**, 2.

21 Jonson, (I 45), 40: Act 2, sc. 1, l. 259.

CHAPTER V

1 It is commonly stated that the earliest printed work ascribed to Basi Valentine was published in 1602: see, for example, E. O. von Lippmann, *Entstehung und Ausbreitung der Alchemie*, 2 vols., Berlin, 1919–1931, **1**, 640, **2**, 53-54; Hoefer, (III 18), **1**, 479; *sub voce* in *Nouvelle Biographie Générale*. The author is indebted to Prof. T. S. Patterson, University of Glasgow, for drawing his attention to a publication of 1599, contained in the Ferguson collection, to which priority must be given. The title-page of this book (which bears the autograph of Johann Thölde) runs as follows:

Ein kurz Summarischer Tractat, *Fratris Basilii Valentini Benedicten* Ordens / Von dem grossen Stein der Uralten / daran so viel tausent Meister anfangs der Welt hero gemacht haben / darinnen das ganze werck nach Philosophischer art für Augen gestalt / mit seiner eigenen Vorrede / für etlich viel Jahren hinterlassen / Und numehr allen Filijs doctrinae zu gutem Publiciret und durch den Druck ans Liecht bracht.

Durch Iohannem Thölden Hessum.   Gedruckt zu Eissleben /
durch Bartholomoeum Hornigk.   Anno M.D.IC.

This work contains the text of the Twelve Keys, without any
illustrative emblems.

2   *Triumph-Wagen des Antimonii* (p. 355), forming pp. 293-446 in *Fr.*
Basilii Valentini Benedictiner Ordens Chymische Schriften (in
2 parts), Hamburg, 1687.

3   Basil Valentine, (I 49), 362.

4   Ferguson, (II 18), **2**, 446.

5   *Idem*, **1**, 81; see also Kopp, (III 28), **1**, 29.

6   Basil Valentine, (I 49), 486.

7   This work has passed through numerous editions in various languages.
The following have been used here: *The Triumphant Charriot of*
Antimony, London, 1660; *Basil Valentine His Triumphant Chariot*
of Antimony, with Annotations of Theodore Kirkringius, London,
1678; *Theodori Kerckringii Commentarius in Currum Triumphalem*
Antimonii Basilii Valentini, Amsterdam, 1685; The Hamburg
edition of 1687 (V 2 above); *The Triumphal Chariot of Antimony*,
London, 1893.

8   Ferguson, (II 18), **2**, 417.

9   See, for example, Thomson, (I 53), **1**, 47.

10 The Triumphant Charriot of Antimony, being A Conscientious Dis-
covery of the many Reall Transcendent Excellencies included in
that Minerall, written by *Basil Valentine* A Benedictine Monke,
Faithfully Englished and published for the Common Good.   By I. H.
Oxon.   Printed for *Thomas Bruster*, and are to be sold at the three
Bibles neere the West end of *Paules Church-Yard* in *London*, 1660.
See p. 97.

11 A. E. Waite (ed.), *The Triumphal Chariot of Antimony, by Basilius*
Valentinus, with the Commentary of Theodore Kerckringius, London,
1893, 23, 24, 25, 26, 55, 82, 36, 37.   This is a translation of the
Latin version, published at Amsterdam in 1685 (see V 7, above).

12 Waite, (V 11), 59.

13 *Idem*, 89.

14 Waite, (I 46), **1**, 20.

15 Waite, (V 11), 175.

16 *Idem*, 41.

17 Hopkins, (I 22), 64.

18 In the preface to *Von den natürichlen und übernatürlichen Dingen*,
Leipzig, 1603.   The edition of the *Twelve Keys* to which Thölde
here refers is presumably that of 1599 (V 1).

19 Basil Valentine, (I 49), 158.

20 ' Oil of vitriol ' is sulphuric acid, but ' vitriol ' is used to designate any
glistening crystalline body.   Thus, copper acetate is a vitriol, prepared as follows (*idem*, 351): ' Take some pounds of verdigreece,
extract its Tincture with distill'd Vinegar, let it shoot [crystallise],
then you have a glorious Vitriol '.

21   *Idem*, 309.   *Cf. Haligraphia* (1603) for such references.

22   The ' Bain-Marie ' was an indispensable adjunct of the alchemist's laboratory, said to have been invented by Mary (Maria or Miriam) the Jewess, the sister of Moses.   The apparatus is described in the *Nomenclator* (1585) as ' a double vessel which being set over another kettle doth boile with the heat thereof seething ', but the term was occasionally applied to a sand-bath as well as to a water-bath.

23   Valentine, (I 49), 302.

24   The Latin version appears also in J. J. Manget's *Bibliotheca Chemica Curiosa*, Geneva, 1702, **2**, 409.

25   Waite, (II 59), **1**, 312.

26   Valentine, (I 49), 221.

27   Waite, (II 59), **1**, 319.

28   Valentine, (I 49), 226.

29   Lewis Carroll, *Through the Looking-Glass*.

30   Valentine, (I 49), 232.

31   Waite, (II 59), **1**, 325 *et seq.*

32   The illustrations of the *Twelve Keys* in the *Viridarium Chymicum* (1624) were printed from the plates used six years earlier for the *Tripus Aureus*.   Most of these engravings measure about 9.5 cm. wide by 7 cm.; nos. VII and IX, however, are about 6.5 cm. square.

33   These and other fanciful alchemical names assigned to antimony— such as ' the red lion ', ' the fiery dragon ', ' the fiery Satan ', ' the son of Satan ', and ' the ultimate judge ',—are given by J. W. Mellor in *A Comprehensive Treatise on Inorganic and Theoretical Chemistry*, London, 1929, **9**, 341.   See also *Proceedings of the Royal Society of Medicine* (Historical Section), London, 1926, **19**, 123.

34   Pernety, (II 53), 545.

35   An earlier print of the emblem of the Twelfth Key, from the plate used in the *Viridarium Chymicum* (1624), appears on the title-page of Mylius' *Basilica Chymica*, published by Lucas Jennis at Frankfurt in 1618.

36   In *The Last Will and Testament*, London, 1671 (I 49 above), the section entitled *A Practick Treatise Together with the XII. Keys and Appendix of the Great Stone of the Ancient Philosophers*, has a separate title-page dated 1670, and the contents begin with page 213.   The Appendix begins on p. 279.

37   A. Stange, *Die Zeitalter der Chemie in Wort und Bild*, Leipzig, 1908, 106.

38   C. J. Sharp, *Folk Songs from Somerset*, London and Taunton, 1911, **4**, 25, 77.

39   Ferguson, (II 18), **1**, 57.

40   The title-page runs as follows: *Le Triomphe Hermetique, ou La Pierre Philosophale Victorieuse*.   Traité Plus complet & plus intelligible, qu'il y en ait eû jusques ici, touchant Le Magistere Hermetique.

A Amsterdam, Chez Henry Wetstein. 1699. A second title-page, following the preface, is headed *L'Ancienne Guerre des Chevaliers.* The curious and complex history of this tract has been partly disentangled by Ferguson—see (II 18), **2**, 486. It is stated in the preface of the French edition here concerned that the work was first issued at Leipzig in 1604: according to Ferguson, as an appendage to Basil Valentine's *Triumphwagen Antimonii,* edited by Thölde. It has appeared under different titles, including *Der Hermetische Triumph* and *Uralter Ritter-Krieg.* An English version is called *The Ancient War of the Knights; or, Victorious Stone.*

Ferguson states that this work has been wrongly attributed to Saint Disdier. An engraving closely similar to the one now described was reproduced by de Givry—(I 5), 360—over the description: 'The Mercurial Hieroglyph: Limojon de Saint-Didier, *Le Triomphe hermétique* (Amsterdam, 1710)'.

CHAPTER VI

1 Holmyard, (II 8), 118.
2 Kopp, (III 28), **1**, 46.
3 D. Diderot, *Encyclopédie, ou Dictionnaire Raisonné des Sciences, des Arts, et Métiers,* 1751–1765.
4 de Givry, (I 5), 355.
5 Artephius, (III 83), 211.
6 This ornamented border was repeated on the title-page of *Gloria Mundi,* in the 1625 edition of *Musaeum Hermeticum.*
7 Measuring about 10 cm. wide by 8 cm.
8 Maier reproduced this alleged engraving of Roger Bacon in his *Tripus Aureus* of the following year (1618) as a delineation of Basil Valentine! See (IV 12).
9 J. B. Craven, *Count Michael Maier: Doctor of Philosophy and of Medicine, Alchemist, Rosicrucian, Mystic. Life and Writings.* Kirkwall, 1910.
10 *Nouvelle Biographie Générale,* Paris, 1863, **32**, col. 862.
11 Craven, (VI 9), 121.
12 Duncan Liddel (1561–1613) graduated M.A. at King's College, Aberdeen, and studied at Rostock (1587–1590) and other German universities. He held many important offices in the University of Helmstedt (1591–1607), and after his return to Scotland endowed a chair of mathematics in the Marischal College, Aberdeen, to which institution he bequeathed his books and instruments. There is a fine contemporary brass (1622) to his memory in St. Nicholas' Kirk, Aberdeen.—See Prof. John Stuart, *A Sketch of the Life of Dr. Duncan Liddel, of Aberdeen,* Aberdeen, 1790; P. J. A. [nderson], *Bibliography of Duncan Liddel,* Aberdeen, 1910.
13 John Johnston was a graduate of King's College, Aberdeen (*c.* 1565–1611), who studied at Rostock (1584), Helmstedt (1585 and

1588–1589), and Heidelberg (1587), and later (1593–1611) held office as Second Master of St. Mary's College in the University of St. Andrews. His works include books published at Leyden in 1607, 1609, and 1612. An eminent Latin poet and scholar, he had a wide acquaintance among the scholars and theologians of his day, and may well have corresponded with Michael Maier, although no definite evidence to this effect has come to light.—See *Fasti Ecclesiae Scoticanae*, 1928, **7**, 428.

Maier's statement in his *Tractatus De Volucri Arborea* (Frankfurt, 1619) that 'lately some of those goslings have been pointed out to me by a certain Scot' suggests that he may even have visited Scotland before returning to Frankfurt from England in September 1616.

14   Kopp, (III 28), **1**, 220.
15   Craven, (VI 9), 8.
16   *Idem*, 68.
17   Waite, (II 103), 318.
18   Much information on this subject is contained in the following work: E. Heron-Allen, *Barnacles in Nature and in Myth*, Oxford, 1928.
19   Waite, (II 30), 58.
20   A. Chalmers, *Biographical Dictionary*, London, 1815, **21**, 138; *Biographie Universelle*, Paris, 1820, **26**, 231; Waite, (II 30), 160.
21   See Horace, *Ars Poetica*, l. 356. In the preceding example, *gerrae* may mean either a wicker-work of osiers, or trifles (nonsense); it is related that at the siege of Syracuse, when the Athenians called for *gerrae* (osiers), the Sicilians mockingly replied ' gerrae ' (nonsense). (The author is indebted to Miss Nancy Dall for these two examples.)
22   Ferguson, (II 18), **2**, 66; Waite, (II 103), 324 (footnote); *Biographie Universelle*, Paris, 1820, **26**, 231.
23   See G. C. Williamson (ed.), *Bryan's Dictionary of Painters and Engravers*, London, 1903, **2**, 21; *Nouvelle Biographie Générale*, Paris, 1855, **7**, 670; the introduction to *Voyages en Virginie et en Floride*, Paris, 1927. The date of Johannes Theodorus de Bry's death is sometimes given as 1623.
24   See *Hariot's Narrative of the First Plantation of Virginia in 1585, printed in 1588 and 1590, reprinted from the edition of 1590 with de Bry's engravings*, London, 1893. The original illustrated edition of 1590 is exceedingly rare, and has fetched as much as £1400 in the auction-room. See also *Voyages en Virginie et en Floride*, with 67 reproductions of de Bry's engravings from originals by Jacques le Moyne (1564) and John White (1585), Paris, 1927.
25   *Shakespeare's England*, Oxford, 1916, **1**, 194.
26   The illustrations of *Secretioris Naturae Secretorum Scrutinium Chymicum* (1687) are usually identical with those of the earlier *Atalanta Fugiens* (1618); occasionally, however, as in the representation of the coral-gatherer (Plate 51), the two prints differ appreciably in the fine detail.

27 See Ovid, *Metamorphoses*, x. 560, *sqq.* It is of interest that tomatoes were originally called, in English, Golden Apples, or Apples of Love. ' Howbeit ', observes Gerard naïvely in his *Herball*, ' there be other golden Apples whereof the Poëts doe fable, growing in the Gardens of the daughters of *Hesperus*.'

28 H. J. Rose, *A Handbook of Greek Mythology*, New York, 1929, 148.

29 Artephius expounds this theme in his *Secret Booke*, (III 83), 193: ' Unlesse the *Bodies* be subtilized and made thinne by *fire* and *water*, untill they doe arise like *spirits*, and bee made like water and fume, or like *Mercury*, there is nothing done in this *Arte*. But when they ascend, they are borne in the ayre . . . therefore it is wisely said that the Stone is borne in the *Ayre*, because it is altogether *spirituall*; for *the vulture flying without wings, crieth upon the top of the mountaine, saying, I am the white of the blacke, and the red of the white, and the Citrine sonne of the red, I tell truth, and lie not*.'

30 Rose, (VI 28), 138.

31 Davis, (I 15), 252.

32 Rose, (VI 28), 115-116.

33 *Idem*, 44.

34 The accompanying German version runs as follows:

Man sagt / dass der Stein sey ein verworffen Ding und schlecht /
    So ligt am Weg / dass ihn Reich und Arm könn haben mit Recht /
Es sagen andre / dass er sey in hohen Bergen zuschawen /
    Oder im Lufft / andre / dass er sich ernehr in Awen /
Alles ist wahr / nach seinem Verstand / Ich wil / dass du warnemmst
Solch Gabe auff Bergen / wo er zufinden bequemst.

35 The accompanying emblem is closely reproduced in the bottom medallion of the title-page of the *Musaeum Hermeticum* (1625)— see Plate 30.

36 Antoine Joseph Pernety (b. 1716) deserves a brief mention, as an interesting but little-known disciple of alchemy. He studied, as a Benedictine, in the abbey of St. Germain-des-Près at Paris. His career was long, varied, and picturesque. In 1763, soon after the publication of his *Dictionnaire* and *Fables devoilées*, he was appointed chaplain to Bougainville's expedition to the Falkland Islands, and later he became librarian to Frederick the Great. At the time of the French Revolution he was living in retirement at Avignon, where he had founded a special sect. His great learning was confused and undigested, and his credulity, like that of Maier, passed all bounds. ' When the Revolution broke out he passed through it as quietly as he could, meddling with nothing, saying nothing, and keeping out of sight; but for all that he passed some months in prison, keener than ever in the search for the Philosopher's Stone, and fully persuaded that he would live for centuries.' He died in 1801 at the age of eighty-five, not thinking, says Thiébault, that he was doing more than falling asleep.— Ferguson, (II 18), **2**, 182.

37   Chaucer, (I 40), 426.
38   For some references bearing upon Greek alchemy and music, see
       J. R. Partington, *Nature*, 1935, **136**, 107.
39   Hoefer, (III 18), **1**, 248-249.
40   *The Merchant of Venice*, act 5, sc. 1, l. 58.
41   Sarton, (II 16), **2**, 25.
42   Waite, (II 30), 50.
43   Ashmole, (I 1), 60.
44   A reproduction of this engraving, in reverse and without the inscrip-
       tions, is given by Lacroix, (I 51), 229.
45   Waite, (II 103), 276.
46   See also H. S. Sleeper, (I 14), 1938, **15**, 410.

CHAPTER VII

1   Details of Stolcius' career are very scanty.   The author is indebted
      for most of the following information to Dr. Josef Volf, Director of
      the Library of the National Museum at Prague.
           The family of Stolz von Stolzenberg, otherwise known as
      Pardubští, came from Kutná Hora (Kuttenberg) in Bohemia,
      formerly celebrated for its silver mines.   Daniel Stolczius Cuttenus
      (von Kuttenberg) took his Bachelor's degree at the University of
      Prague on August 9th, 1618, during the rectorate of Fradelius, the
      theme which he affirmed being entitled *An bella et intestina ex
      astris praevideri possint* ('Whether wars, including civil wars, can
      be foretold from the stars ').   His theme for the Master's degree,
      to which he was promoted on October 3rd, 1619, was also of an
      astrological nature: *Utrum sidera prospero motu in aethere moveantur*
      ('Whether constellations move favourably in the sky ').
           On May 29th, 1621, Stolcius appears on the matriculation roll
      of the University of Marburg, under the name, M. Daniel Stolcius
      Cuttenbergenus.   Although Stolcius dated the dedication of his
      Viridarium Chymicum from Oxford, 6/16 July 1623, he does not
      appear to have been admitted as a member of Oxford University.
2   *Viridarium Chymicum Figuris Cupro incisis adornatum, et Poeticis
      picturis illustratum:* . . . Authore M. Daniele Stolcio de Stolcen-
      berg Bohemo Med. Candidato.   Francofurti Sumptibus Lucae
      Jennisi.   1624.
           The original Latin (1624) and German (1624) editions of this
      work are very rare.   It is of interest that the Bodleian Library,
      Oxford, contains neither, and the Library of the Národní Museum,
      Prague, has only the German edition.
3   *Tripus Aureus*, Frankfurt, 1618.   The two illustrations following the
      emblems of the Twelve Keys in the *Viridarium* occur on p. 67 (see
      Fig. 13) and the title-page (see Plate 32), respectively.
4   From the title-page of *Lusus Serius*, Frankfurt, 1616.   Probably
      engraved by J. T. de Bry.   Figura XV of Stolcius.

5   Waite, (II 103), 317.
6   Numbered XVI to XXVII by Stolcius.
7   Stolcius, Figura XVII.
8   *Idem*, XX.
9   Artephius, (III 83), 226.
10   Stolcius, Figura XXV.
11   These engravings, measuring about 10 cm. wide by 7.5 cm., are numbered XXVIII to XXXII by Stolcius.
12   Homer, *Odyssey*, x, 302-306.
13   Numbered XXXIII to XCIII by Stolcius.
14   Ioannis Danielis Mylii, T. & Med. Candidati Wetterani Hassi, *Philosophia Reformata* Continens Libros binos. Francofurti apud Lucam Iennis, Anno M.DC.XXII. The title-page design bears the name Baltzer Schwan.
15   Measuring about 7 cm. wide by 6 cm.
16   One of these illustrations is reproduced twice, under the numbers LXI and LXXXI (see Note 29 below). The engravings in Mylius' *Philosophia Reformata* are arranged in three series, containing 28, 20, and 13, respectively. The number of each engraving appears in the top left-hand corner. Twenty-four of them, together with five other designs in Stolcius' collection, were printed yet again from the same plates, in 1625, in another work published by Lucas Jennis at Frankfurt; this is an allegorical poem dealing with the preparation of the Philosopher's Stone, and written by Herbrandt Jamsthaler under the title *Viatorium Spagyricum* ('Spagyric Guide'). Seven of them, also, were used to illustrate the versified German edition of Norton's *Ordinall of Alchimy*, or *Crede-Mihi*, published by Lucas Jennis at Frankfurt in 1625.
17   Stolcius, Figura XXXIII; Mylius, 1.
18   Stolcius, XXXIV; Mylius, 2.
19   Stolcius, XXXV; Mylius, 3.
20   Stolcius, XXXVI; Mylius, 4.
21   These drawings are numbered as follows by Stolcius and Mylius, respectively: calcination, XXXVII (5); solution, XXXVIII (6); separation, XXXIX (7); conjunction, XL (8); putrefaction, XLI (9); congelation, XLIII (11); cibation, XLIV (12); sublimation, XLV (13)—see Plate 56 (i); fermentation, XLVI (14); exaltation, XLVII (15); multiplication, XLVIII (16)—see Plate 56 (ii).
22   Stolcius, Figura XLII; Mylius, 10.
23   The 32 engravings are numbered as follows by Stolcius and Mylius, respectively: XLIX (17) to LX (28), and LXI (1) to LXXX (20). Owing to an error, Stolcius interchanged (25) and (28) in Mylius' first series, so that these are numbered LVII (28) and LX (25).
24   Stolcius, Figura LXXVI; Mylius, 16.
25   Stolcius, LXXX; Mylius, 20.
26   Stolcius, LV; Mylius, 23.
27   Stolcius, L; Mylius, 18.

28   Stolcius, LXIV; Mylius, 4.

29   The Epilogue to Book I occupies pp. 312-384 of *Philosophia Reformata*. The large illustration on p. 316, measuring about 11 cm. wide by 11.6 cm., showing two men conversing beneath the tree of universal matter (Plate 58), is printed again on the title-page of Book II (p. 365); it is the only illustration in Book II. This illustration forms No. 1 in Mylius' third and last series (of 13). Owing to the large size of the plate, Stolcius replaced it in his collection by reproducing the first drawing of Mylius' second series, so that LXXXI and LXI are identical in the *Viridarium*. The other twelve drawings in the Epilogue are numbered as follows by Stolcius and Mylius, respectively: LXXXII (2) to XCIII (13).

30   These 13 engravings are numbered XCIV to CVI in Stolcius' collection. The final engraving of the *Viridarium* (CVII), entitled 'Insignia of our Hero', is a variant of the first illustration of *Splendor Solis* (see p. 67).

31   Two qualifications must be made here. Firstly, Stolcius LXXXI should really show the two men conversing beneath the tree of universal matter, as explained in Note 29 above: it then corresponds to CV (Plate 62). Secondly, XCII ('Dream or Vision', referring to the introductory section of the Epilogue) in the first series, and XCV (the 'vitriol acrostic' of Basil Valentine) in the second series, must be omitted from the comparison. The remaining 11 pairs of illustrations correspond exactly, thus: LXXXII-XCIV, LXXXIII-XCVI, LXXXIV-XCVII . . . XCI-CIV, and XCIII-CVI.

32   This tract, in Latin, is summarised on the title-page as follows: 'Azoth or the Hidden Gold-making Secrets of the Philosophers, explaining concisely and clearly the First Matter and that renowned Stone of the Philosophers to the Sons of Hermes: By means of the Riddle of the Philosophers, speaking in Parables; the Emerald Table of Hermes; Symbols, Parables and Designs of Saturn, of Brother Basil Valentine.' Johann Bringer also published a German edition of the same tract, in 1613—see Ferguson (II 18), **2**, 150.

    Azoth is sometimes ascribed to Basilius—see, *e.g.*, Kopp, (III 28), **1**, 30—although the only mention of this mysterious writer which it contains, beyond the above reference on the title-page, is to be found in a heading 'Symbolum F. Basilii Valentini', or 'Symbol of Brother Basil Valentine', given to a short paragraph on some of the mystic aspects of the Stone. *Azoth*, through its appeal to the superstitious credence of would-be gold-makers, found publication in several forms during the seventeenth century —see *e.g.* Ferguson, (II 18), **1**, 57; **2**, 150—and in some of these it was definitely ascribed to Basilius, presumably in order to increase the sales.

33   The word 'Azoth' was endowed with a cabbalistic significance by alchemists, as it contains the first and last letters, Aleph and Thau

of the Hebrews, Alpha and Omega of the Greeks, and A and Z of the Romans. It was applied to a large variety of imaginary bodies, principles, and remedies. In the literature of the Philosopher's Stone it was often used as a synonym for sophic mercury. See, for example, *Philosophia Reformata*, Frankfurt, 1622, 362; *Azoth*, Paris, 1659, 175; Pernety, (II 53), 52; Kopp, (III 28), **1**, 30 (footnote).

34   The whole of this Epilogue, with the introduction, was reprinted in J. J. Manget's *Bibliotheca Chemica Curiosa* (Geneva, 1702, **2**, 198), with an added note that ' the primal matter is clearly and manifestly explained thereby ' and that ' The Riddle of the Philosophers, or Symbol of Saturn, by means of parables clearly sets forth Azoth '.

A close French translation of the Epilogue, without the introduction, was published in 1659 under the catchpenny title, ' Azoth, or the Way to Make the hidden Gold of the Philosophers '. In this edition, the authorship is boldly ascribed to Basil Valentine on the title-page, which is ornamented with a simplified version of Mylius' tree of universal matter. The same illustration occurs on the title-pages of the 1613 editions of *Azoth* (see Note 32 above), with the date 1605 at the foot (Fig. 17). The title-page runs as follows (the original French edition appeared in 1624):

Azoth, ou le Moyen de Faire l'Or caché des Philosophes. *De Frere Basile Valentin.* Reueu, corrigé & augmenté par Mr. L'agneau Medicin. A Paris. Chez Pierre Moet, Libraire Iuré, proche le Pont S. Michel à l'Image S. Alexis. M.CD.LIX.

The symbolic drawings contained in the second part of this French edition of *Azoth* are inferior copies of the illustrations in the Frankfurt editions of 1613. The reproductions given in the present work have been taken from the superior set which form the last section of Stolcius' *Viridarium*; these vary from about 6 to 7.5 cm. wide by 6.5 to 7.5 cm.

35   The same engraving is used as a frontispiece to *Gloria Mundi*, in the *Musaeum Hermeticum* of 1625, also published by Lucas Jennis at Frankfurt.

36   Stolcius, Figura XCIV.

37   *Idem*, XCV.

38   *Idem*, XCVI.

39   *Idem*, XCVIII.

40   *Idem*, XCIX.

41   *Idem*, CI.

42   *Idem*, LXXXVIII.

43   *Idem*, CV.

44   Pernety, (II 53), 522.

45   The same emblem appears also on the title-page of a short treatise included in *The Hermetic Museum*, under the name: ' A very brief tract concerning the Philosophical Stone, written by an unknown German Sage, about 200 years ago, and called the Book of Alze,

but now published for the first time '.   See *Musaeum Hermeticum*, Frankfurt, 1625, 429; also Waite, (II 59), **1**, 259.

46   Ferguson is mistaken in stating that this second *Viridarium Chymicum* of 1688 contains the twelve keys of Basilius and the twelve emblems from *Symbola Aureae Mensae*, together with 28 emblems from *Atalanta Fugiens* and 8 modified reproductions from the *Rosarium Philosophorum*.   These last two works are not concerned in Herman von Sand's publication.   See Ferguson (II 18), **2**, 411. The copy in the James Young collection, mentioned by Ferguson, is similar to that in the British Museum (B.M. Library, 1035 a 32).

A note may be added on the last of the works mentioned by Ferguson, as it had a great vogue among alchemists: *Rosarium Philosophorum*, Francofurti, ex off. Cyriaci Jacobi, 1550.   This rare editio princeps contains 96 leaves, without pagination, and is illustrated with 21 large woodcuts, some of which would not commend themselves to modern taste.   An old MS. note on the title of the St. Andrews copy reads thus: ' Georgius Anrac, vel Aurac, et Lansac Auctore 1470 '.

47   D. Stolcio de Stolcenberg, *Hortulus Hermeticus Flosculis Philosophorum*, Francofurti, Impensis Lucae Jennisii, 1627.   See Lenglet Dufresnoy, (I 6), **3**, 300; Ferguson, (II 18), **2**, 410.

48   The engravings appeared originally in the third division of Mylius' *Opus Medico-Chymicum*, published by Lucas Jennis at Frankfurt in 1618.   This work also contains Fig. 8, p. 86.

49   Kopp, (III 28), **1**, 207.

50   de Givry, (I 5), 352.

The Crowned Salamander.   From *Bibliotheca Chemica Curiosa*, Manget, 1702.

INDEX

(See also the summary on pp. ix-xi)

319

INDEX

INDEX

Hesperian Garden, 169, 196, 221, 237, 239, 259
Hesse, 268
Hieroglyphics, alchemical, 90, 245
Hiram, 259
Hippomenes, 237
Histoire de la Philosophie Hermétique, 3
Hoefer, 1, 21, 248
Holmyard, 8, 41, 53, 213
Holy Water, 14
Homer, 79
Hoover, 30, 78
Hopkins, 11-16, 18, 24, 44, 108
Hormones, 142
Horseshoe, 88
Hortulanus, 52, 227
Horus, 21
Hsien, 122
Huan-tan, 122, 145
Humboldt, 43
Hunting of the Green Lyon, 87
Hylozoism, 12, 94
Hypostatical principles, 27, 32

Iatro-chemistry, 29, 30, 44
Ibn Sina, 17
Imbibition, 139
Imhotep, 5
Imitation gold, 13, 40, 146
Incombustible oil, 103, 160, 199
India, 4, 8, 9, 19, 101
Infants' blood, 63, 158, 266
Inhumation, 139
Iris, 259
Iron, 88
Isaac of Holland, 123, 134
Isis, 21, 39, 103
Islamic era, 17, 21

Jabir, 17, 33, 53
Jacob's ladder, 156
Janitor Pansophus, 84, 167, 168
Jason, 87, 163, 275
Jean de Ré, 47
Jennis, Lucas, 166, 169, 174, 222, 233, 255-257, 259, 260, 267, 273, 274
Johnston, 230, 311
Jonson, Ben, 25, 36, 39, 63, 93, 103, 104, 124, 129, 135, 136, 145, 161, 163, 212, 223, 259
Jove, 88
Jud he voph hé, 128
Juniper, 172
Juno, 161

Jupiter, 88, 89, 198, 285; *see also* Zeus

Kekulé, 1, 241
Kelly, 227
Kepler, 209
Kerotakis, 15, 40, 55
Khunrath, 81, 82, 87, 251
King, 102, 151, 184, 200, 262, 269, 271
Kobolds, 79
Ko Hung, 6, 38, 145
Kopp, 69, 184
Kronos, 63, 88, 143, 158, 201, 243; *see also* Saturn
Kuhdorfer, 214

Laborant, 71
Laboratories, 71, 77, 81, 179, 213, 251, 274, 275, 307
Lacinius, 56
Lacroix, 28, 30, 60, 74, 113
Lac virginis, *see* Virgin's milk
Ladon, 237
Langmuir, 109
Lao-Tsŭ, 5
Last Will and Testament, 90, 184, 193 *et seq.*
Lato or Latona, 93, 103, 161, 165, 218, 240, 262
Laurel, 260
Lavoisier, 31, 33
Lead, 12, 88
Lenglet Dufresnoy, 3
Leto, *see* Lato
Leyden papyrus, 13, 38
Libavius, 30, 80, 213 *et seq.*, 266
Liddel, 229, 311
Liebig, 1, 93, 118, 119, 121, 164
Lili, 146
Lily, 201, 216, 219
Lion, 84, 92, 206, 219, 240, 262, 264, 265, 271
Literature of alchemy, 36 *et seq.*
Little mountains, 150
Loadstone, 97
Lockyer, 107
Logos, 109
Longevity, 6, 122
Lu-Chiang Wu, 7
Lucina, 256
Lully, Raymond, 43, 47, 51, 56, 116, 170, 223
Luna, 83, 88, 102, 157 *et seq.*, 167, 198, 204, 208, 240, 271

INDEX

Oil of vitriol, 194
Oils, 4, 50, 191
Old Moore, 213
Olive, 259
Olympian deities, 88, 265
Olympiodorus, 40, 41
Operations of the Great Work, 83, 262
Opus Majus, 43
Opus Saturni, 123
Ordinall of Alchimy, 85, 99, 101, 130, 169, 174 *et seq.*, 250, 253
Organic chemistry, 192
Orion, 243
Osiris, 20, 33, 103
Ouroboros serpent, 108, 117
Ovid, 68, 239
Owl, 81, 167
Oxford, 255, 256, 258
Oxygen, 33, 186

Palestine, 8
Palsy, 77
Pao P'u Tzŭ, 6, 7
Papyrus Ebers, 37
Paracelsus, 27, 29, 30, 44, 49, 81, 82, 126, 137, 148, 164, 184, 186
Paris, 22, 59, 65, 174
Partington, 8
Parvati, 19
Patriarchs, 124
Peacock, 93, 146, 202, 216, 219, 262
Pelagius, 41
Pelican, 149, 167, 211, 264, 265
Penelope, 161
Pernety, 120, 128, 136, 139, 140, 160, 205, 235, 245, 246, 272, 313
Perrenelle, 64 *et seq.*, 101, 124
Persia, 8
Persian texts, 41
Phallism, 85, 106
Philalethes, 47, 92, 129, 136, 146, 150, 154, 160, 167, 248
Philosopher's Dunghill, 146
Philosopher's Egg, 41, 104, 146, 181, 208, 217, 239, 241, 270; *see also* Vase of Hermes
Philosophers' Heaven, 299
Philosopher's Stone, 2, 12, 17, 54, 60, 68, 69, 83, 84, 94, 101, 107, 110, 112, 118 *et seq.*, 202, 204, 207, 214, 238, 244, 248, 258, 265, 275, 285
Philosophia Reformata, 260 *et seq.*
Philosophical mercury, 25
Philosophical sulphur, 25

Philosophic Tree, *see* Tree of the Metals
Philosophorum Praeclara Monita, 47, 63
Philosophus, 268
Phlogiston, 31 *et seq.*, 144
Phoebus, 163, 260
Phoenix, 84, 167, 199, 204, 219
Pill of immortality, 122
Pinguedo terrae, 33
Planets, 17, 20, 83, 84, 88, 89, 95, 99, 104, 153, 204, 265
Plato, 9, 11, 108, 247
Plautus, 223
Pleasure Garden of Chymistry, 256 *et seq.*
Plotinus, 9
Plutarch, 240
Poisons, 108, 189
Poisson, 28, 29
Polygraphice, 1
Practical tradition in alchemy, 74 *et seq.*
Prague, 22, 229, 255
Prayer, 81, 83, 159, 251, 277, 278
Precepts of Hermes, 54
Precious stones, 125
Prescriptions, 37, 77, 189
Pretiosa Margarita Novella, 48, 51, 55-57, 265
Price, 142
Priestley, 33
Prima materia, 10, 26, 95, 120, 168
Primitive materials, 131, 202
Printing, 35, 92
Processes of the Great Work, 136 *et seq.*
Projection, 66, 141, 264
Proximate materials, 131
Pryme, Abraham de la, 173
Pseudo-alchemists, 2, 13, 23, 28, 70; *see also* Puffers
Pseudo-Aristotle, 19
Pseudo-Avicenna, 19
Pseudo-Democritus, 40, 193
Pseudo-Lully, 142, 148, 186
Pseudo-Roger Bacon, 24, 142
Psyche, 160
Puffers, 2, 23, 92, 126, 134, 161, 203, 252
Putrefaction, 95, 138, 147, 203, 204, 216, 218, 263
Pyrgopolynices, 222, 223, 225
Pythagoras, 11, 29, 37, 56, 209, 247
Python, 161, 162, 165, 262

INDEX

Hie endet sich diß buoch
seligklich getruckt vnd volendet in der Kei-
serlichen stat Straßburg durch Joha̅
nem Grüninger vff Sa̅t Adolffs
abent In dem Jar so ma̅ zalt
nach Christi geburt. M.ccccc
vnd.xix.

The End of the Book.   Brunschwick, 1519.